MULTIPLE CHOICE QUESTIONS IN HUMAN PHYSIOLOGY

Multiple Choice Questions in Human Physiology

With Answers and Comments

IAN C. RODDIE
D.Sc., M.D., F.R.C.P.I.

Dunville Professor of Physiology, The Queen's University of Belfast, and Consultant in Physiology to the Northern Ireland Eastern Health & Social Services Board

WILLIAM F. M. WALLACE
B.Sc., M.D., F.R.C.P.

Professor of Applied Physiology, The Queen's University of Belfast, and Consultant in Physiology to the Northern Ireland Eastern Health & Social Services Board

Third Edition

Edward Arnold
A division of Hodder & Stoughton
LONDON MELBOURNE AUCKLAND

First published in Great Britain 1971
Second edition 1977
Third edition 1984
Paperback version 1989
Reprinted 1991, 1992

(This work has also been translated into the Italian and Japanese languages)

British Library Cataloguing in Publication Data

Roddie, Ian (Ian Campbell)
Multiple choice questions in human physiology – 3rd ed.
1. Physiology. Questions and answers
I. Title II. Wallace, William F. M. (William Frederick Matthew)
612′.0076

ISBN 0-7131-4588-9

Printed and bound in Great Britain for Edward Arnold, a division of Hodder and Stoughton Limited, Mill Road, Dunton Green, Sevenoaks, Kent TN13 2YA by Clays Ltd, St Ives plc.

CONTENTS

PREFACE

In this book our aim has been to base the questions on those aspects of Physiology which have most relevance to medical practice and enjoy wide general acceptance. Thus, it is hoped that the questions will be useful to medical, dental and science undergraduates taking courses in human physiology and to postgraduate medical students reading for higher examinations. We have tried to avoid excessive detail in the way of facts and figures; in general those which are included are of value in medical practice.

In preparing for this third edition, we have again thoroughly reviewed all questions. About half have been modified—many to a minor extent—to increase their clarity and usefulness. Four questions have been deleted and 61 new questions added. In particular, questions involving the use of diagrams, introduced in the second edition, have been doubled in number to nearly 10% of the total. We feel that such questions are particularly useful in testing high levels of understanding and manipulation of physiological concepts and data.

Each question consists of a common stem giving rise to four branches. The stem and a single branch together constitute an independent statement to be judged "true" or "false" by the reader. Care has been taken that the statements in any question are not mutually exclusive. Therefore, four independent decisions are required to answer each question. This system has the advantage of simplicity and brevity over most other forms of multiple choice question.

The questions are grouped in ten sections to cover the various physiological systems. Each section is divided into two parts, *basic physiology* and *applied physiology*. *Applied* questions require an elementary knowledge of medical problems but are designed so that the answers may be deduced mainly by making use of basic physiological knowledge.

We are indebted to colleagues for suggesting questions and to all who have commented on previous editions. We would also like to pay tribute to the consistent courtesy and professional skill of our publisher Mr Douglas Luke.

September, 1983

I.C.R.
W.F.M.W.

HOW TO USE THE BOOK

A stimulus to fill gaps in your knowledge

The purpose of this book is to help you to revise your Physiology, to prepare for examinations and to use physiological concepts in your continued self-education at undergraduate and postgraduate level. It has been our experience that students who understand the Physiology they learn have little difficulty in remembering it, whereas those who learn in a parrot fashion are unlikely to do well in examinations or to make good use of their knowledge subsequently. The problem is that some students think they understand a subject as they read it, though their subsequent performance shows that this has not been so. The statements in this book are presented so that you can test your understanding while revising from the notes or textbook of your choice. By committing yourself to an opinion during revision you have an opportunity to confirm correct impressions and when mistaken, to pursue the matter until you understand why you went wrong. Comments are given with most answers to reinforce the correct answer.

We do not pretend that Physiology is merely a collection of true or false statements, or that the statements in this book are all 100 per cent true or false for all time. However, we hope that you will find on investigation and reflection, that most of the statements in our book can reasonably be classified as true or false.

Using the masking card

We suggest you use a piece of plain card, wide enough to cover the columns and print on the right hand page.

1. Open the book at the desired section and *immediately* cover the upper answers on the right-hand page with the card.
2. Slide card to right until the 3 columns on right-hand page are just exposed.
3. Mark in the right hand of the 3 columns T (True) or F (False) for each part of each question. *Committing yourself to a written verdict is an important part of the exercise.* All, some or none of the 4 parts of each question may be true. It is important to remember that every word counts in a multiple choice question. Questions which appear to be "catches" should fail to catch if this is remembered.
4. Slide card downwards to expose answers and comments for first question.
5. When answering the questions for the second time slide card to right until

only 2 columns are exposed. Note particularly those questions which you have answered wrongly on both occasions. You may wish to answer the questions and check the answers one part at a time when first using the book and answer and check a complete question at a time subsequently.

6. Always try to understand why you were wrong (or right) and, if necessary, refer to a textbook or your tutor. If you feel, after due consideration, that a question is confusingly worded or that the answer is wrong, the authors would be most grateful if you would write to them about it.

Scoring your answers

The main purpose of this book is to assist understanding of Physiology, but if you wish to assess your progress, the following method may be applied.

1. Answer 25 questions (4 items per question) or the equivalent by marking them "T" or "F". Do not refer to the answers and comments until you have completed all 25 questions. If you have no idea of the answer (so that tossing a coin would literally be your only way of reaching a decision), leave a blank. If you have some idea of the subject you are, in general, more likely to guess correctly than wrongly. A reasonable time for the 25 questions is 50 minutes.
2. Score 1 for each correct answer.
   – 1 for each incorrect answer.
   0 for each blank.

   The negative marks for incorrect answers are necessary to allow for the fact that random guessing of the answers will result in approximately equal numbers of correct and incorrect answers.
3. The following is a very approximate assessment of your knowledge based on the authors' experience with medical, dental and science students. Naturally this assessment refers to your first attempt only.

50–60 fair
60–70 good
70–90 excellent
90–100 outstanding

BODY FLUIDS
BASIC PHYSIOLOGY

1. Extracellular fluid differs from intracellular fluid in man in that:
(a) it has a greater volume
(b) it has a lower tonicity
(c) its anions are mainly inorganic
(d) it has a higher sodium: potassium molar ratio.

2. Blood group antigens are:
(a) attached to the haemoglobin molecule
(b) inherited characteristics carried by the autosomes
(c) inherited as recessive characteristics
(d) sometimes found in tissues other than blood.

3. Total body water:
(a) forms a smaller percentage of body weight in fat persons than in thin
(b) can be measured by an indicator dilution technique using deuterium oxide as the indicator
(c) comprises half to two-thirds of body weight in young adults
(d) is a smaller proportion of body weight in men than in women.

4. The breakdown of erythrocytes in the body yields:
(a) iron, most of which is excreted in the urine
(b) the pigment bilirubin which is conjugated in the liver to bilirubin glucuronide
(c) amino acids which are available for general use
(d) erythropoietin, thereby regulating the further production of erythrocytes.

5. If a person has group B blood:
(a) he may have the genotype AB
(b) his father might have group O blood
(c) his children will have either group B or group O blood
(d) and his wife also has blood group B, all their children will have blood group B or O.

1.
(a) F cells contain half to two-thirds of body water
(b) F if this were the case, water would be drawn by osmosis into the cells
(c) T the principal extracellular anions are chloride and bicarbonate; inside cells the principal anions are proteins and organic phosphates
(d) T extracellular ratio around 30:1; intracellular about 1:10.

2.
(a) F they are part of the red cell membrane
(b) T
(c) F they are mendelian dominants
(d) T they are sometimes present in saliva.

3.
(a) T fat contains relatively little water compared with other tissues
(b) T deuterium oxide (heavy water) can exchange with total body water
(c) T the percentage tends to fall with age
(d) F women, by carrying relatively more fat, have a lower average proportion of water than men.

4.
(a) F most of the iron is retained for further use
(b) T the conjugated bilirubin is then excreted in the bile
(c) T from the globin portion of haemoglobin
(d) F erythropoietin is not a breakdown product of erythrocytes; its formation is related to the oxygen carrying capacity of the blood (detected by the kidney).

5.
(a) F if this were the case he would be group AB
(b) T in which case he would have inherited O from his father and B from his mother to have the genotype BO
(c) F it depends on the mother's contribution
(d) T both parents, if homozygous, will contribute B or if heterozygous, will contribute B or O.

6. Blood platelets are important in stopping bleeding in that:

(a) they can release lipids which have a role in the clotting process
(b) the bleeding time tends to increase when the platelet count is low
(c) on exposure to collagen they become sticky and adhere together
(d) a fall in the platelet count prolongs the clotting time before the bleeding time is affected.

7. Bilirubin:

(a) is a steroid pigment
(b) travels from red cell breakdown sites to the liver dissolved in plasma in an unbound form
(c) contains iron
(d) is necessary for the digestion and absorption of fat in the intestines.

8. Monocytes:

(a) originate from precursor cells in the lymph nodes
(b) are phagocytic
(c) may migrate from the blood into the tissues
(d) can manufacture certain immunoglobulins (IgM).

9. Erythrocytes:

(a) are responsible for the major part of blood viscosity
(b) contain carbonic anhydrase
(c) generate energy from glucose so that oxygen may combine with haemoglobin
(d) swell up and burst when suspended in hypertonic saline (e.g. three times normal).

10. Human plasma albumin:

(a) makes a greater contribution to the colloid osmotic pressure of the plasma than does globulin
(b) is freely filtered at the renal glomerulus
(c) behaves as an anion at the pH of blood
(d) is involved in the carriage of carbon dioxide in the blood.

6.
(a) T this is part of the intrinsic blood clotting process
(b) T due to impaired platelet plugging
(c) T vascular leaks may be thus sealed by platelet plugs
(d) F very few platelets are required to initiate the clotting process.

7.
(a) F it is a porphyrin pigment derived from haem
(b) F it is bound to plasma protein
(c) F iron has been removed from the haem in the formation of bilirubin
(d) F the bile salts, not the bile pigments are involved in fat digestion and absorption.

8.
(a) F they originate from precursor cells in the bone marrow
(b) T they can ingest dead granulocytes
(c) T where they may become scavenger cells (macrophages) in the connective tissue
(d) F immunoglobulins are manufactured by the lymphocyte series.

9.
(a) T blood viscosity increases exponentially as the haematocrit rises
(b) T it catalyses the reaction $H_2O+CO_2 \rightleftharpoons H_2CO_3$
(c) F energy is generated from glycolysis to maintain the electrochemical gradients across their membranes
(d) F they shrink (crenate) in hypertonic saline; they swell in hypotonic saline (e.g. one-third normal).

10.
(a) T firstly there is more albumin, secondly it has a lower molecular weight
(b) F only a small proportion is filtered and this is reabsorbed by the tubular cells
(c) T the isoelectric point for albumin is on the acid side of neutrality so that negative COO^- groups predominate at blood pH (7.4)
(d) T as carbamino compound: $-NH_2+CO_2 \rightleftharpoons NH.COOH$.

11. Neutrophil granulocytes:

(a) are the most numerous leucocytes in blood
(b) have a life span of about 120 days
(c) are formed mainly in the spleen
(d) contain actin and myosin filaments.

12. Bleeding from a small cut in the skin:

(a) is normally diminished by local vascular spasm
(b) ceases within about 5 minutes in normal people
(c) will be prolonged if factor VIII (antihaemophilic globulin) is absent
(d) is likely to be greater if the skin is warm than if it is cold.

13. Antibodies:

(a) are all proteins
(b) are not formed in response to exposure to antigen in early fetal life
(c) tend to be produced in larger quantities in response to the second than in response to the first exposure to antigen
(d) which circulate as free immunoglobulins are produced mainly by the B lymphocyte series.

14. Red cells in human peripheral blood vessels:

(a) include about 1% from which the nucleus has not been extruded
(b) include about 1% reticulocytes, i.e. cells which have a reticular pattern when appropriately stained
(c) are distributed randomly in the stream of blood
(d) travel at a slower velocity through venules than through capillaries.

15. Lymphocytes:

(a) form 1–2% of the white cell count
(b) are motile
(c) can be transformed by a suitable stimulus into plasma cells
(d) can not cross the capillary wall.

11.
(a) T they comprise 60–70% of circulating leucocytes
(b) F they survive for 1–2 weeks at the most
(c) F they are formed in the bone marrow from undifferentiated cells
(d) T this may be related to their motility.

12.
(a) T vascular smooth muscle contracts because of local damage; serotonin from platelets may contribute to vasoconstriction
(b) T this is the upper limit of the normal "bleeding time"
(c) F bleeding time is independent of clotting time (which will be increased in this case)
(d) T since warmth dilates cutaneous blood vessels.

13.
(a) T manufactured by the ribosomes in plasma cells
(b) T this *immunological tolerance* prevents the fetus forming antibodies which would react with its own proteins
(c) T this phenomenon is known as sensitisation
(d) T the cellular immunity which is responsible for graft rejection is produced mainly by the T lymphocyte series.

14.
(a) F nucleated red blood cells are not seen in normal peripheral blood
(b) T these are the most immature cells normally found in peripheral blood
(c) F they tend to move away from the wall to form an axial stream
(d) F flow is slower in capillaries because their total cross-sectional area is greater.

15.
(a) F 20–40%
(b) T in tissue culture they can throw out pseudopodia which can enter other cells
(c) T it is in this form that they produce antibody
(d) F they can migrate from the blood to the lymphatic capillaries and recirculate back to the blood.

16. The specific gravity of:
(a) red cells is less than that of the plasma
(b) plasma is related more to its protein content than to its electrolyte content
(c) plasma decreases as extracellular fluid volume decreases due to loss of water and electrolytes
(d) blood is higher on average in women than in men.

17. Blood:
(a) comprises about 7% of body weight
(b) comprises a higher percentage of body weight in fat people than in thin people
(c) volume may be calculated if the plasma volume and the haemoglobin concentration are known
(d) volume tends to rise when water is drunk.

18. Human cell membranes:
(a) are more permeable to sodium (atomic weight 23) than to potassium (atomic weight 39)
(b) are impermeable to fat-soluble substances
(c) in skeletal muscle have their permeability to glucose altered in the presence of insulin
(d) consist almost entirely of protein molecules.

19. The viscosity of the blood:
(a) appears greater when measured in the perfused limb (*in vivo*) than when measured in a glass tube (*in vitro*)
(b) at 20°C is less than at 37°C
(c) in veins is greater than that in arteries
(d) increases if the plasma protein content increases.

16.
(a) F red cells are heavier and hence tend to sediment
(b) T plasma proteins (70–80 g/litre) far exceed in weight the electrolytes (about 10 g/litre)
(c) F the reverse is true; plasma specific gravity is an indicator of extracellular volume if the protein level is normal
(d) F it is higher in men because they have a higher haematocrit.

17.
(a) T e.g. 5 kg (approximately 5 litres) in a 70 kg man
(b) F the reverse is true, since fat tissue is relatively non-vascular
(c) F it is necessary to know the plasma volume and the haematocrit
(d) T the water is absorbed into the blood and reduces its osmolality.

18.
(a) F the reverse is true, perhaps because the hydrated sodium molecule is larger than the hydrated potassium molecule
(b) F fat-soluble substances can penetrate cell membranes readily
(c) T insulin increases muscle cell membrane permeability to glucose, potassium and other substances
(d) F carbohydrate and phospholipid are important constituents.

19.
(a) F the *in vivo* value is considerably lower
(b) F it is greater at 20°C (e.g. in cold fingers)
(c) T the venous haematocrit is higher than the arterial because some of the fluid is lost from the capillaries and returned by lymphatics
(d) T but the contribution of plasma proteins is much less important than that of the red cells.

20. Neutrophil granulocytes:
(a) are motile
(b) are confined to the circulation
(c) contain lysosomes
(d) are attracted chemically to sites of inflammation.

21. Cerebrospinal fluid:
(a) is actively secreted by the choroid plexuses
(b) is the major source of the brain's nutrition
(c) has the same pH as arterial blood
(d) is virtually glucose free.

22. Red cells:
(a) are present in larger numbers per unit volume of blood in men than in women
(b) in an individual with a red cell count of 5×10^{12}/litre and a haematocrit of 0.45 have a mean volume of 225 fl (1 fl, femtolitre = 1 μm^3)
(c) in the above individual have a mean haemoglobin concentration of 33% if the blood haemoglobin level is 15 g/100 ml
(d) sediment at a rate which is related to their tendency to come together in rouleaux.

23. Antigens:
(a) are always proteins
(b) in some cases do not produce a complete response unless a plasma factor "complement" is present
(c) produce less immune response than normal in animals which have had the thymus removed at birth
(d) produce a larger antibody response when given as a single whole dose than in two half doses separated by an interval of several weeks.

20.

(a) T they move with amoeboid movement

(b) F they insinuate their way out of capillary blood vessels by the process of diapedesis

(c) T the granules are modified lysosomes which contain the acid hydrolases which digest phagocytosed material

(d) T this process is called chemotaxis.

21.

(a) T secretion is thought to take place since the composition of CSF differs somewhat from that of protein-free plasma

(b) F brain nutrition is catered for by cerebral blood flow

(c) F about 7.3 as compared with 7.4

(d) F its glucose concentration is around 60% of the plasma value.

22.

(a) T men average 5.5×10^{12}/litre; women 5×10^{12} litre

(b) F mean cell volume

$$= \frac{\text{red cell volume/litre}}{\text{red cell count/litre}}$$

$$\text{e.g. } \frac{0.45}{5\times10^{12}}\text{ litres} = \frac{0.45\times10^{15}}{5\times10^{12}} = 90 \text{ fl}$$

(c) T volume of cells containing 15 g Hb = 45 ml; so the Hb concentration within them = 15/45 g/ml = 33%

(d) T the tendency is related to the pattern of proteins in the plasma.

23.

(a) F often, but not always; e.g. some large carbohydrate molecules are antigenic

(b) T e.g. in some antigen-antibody reactions involving red cells complement must be taken up before lysis of the cells results

(c) T the thymus is important in development of the immune response after birth

(d) F the first dose "sensitises" the recipient to the antigen. The second acts as a "booster" as in immunisation.

24. Eosinophil granulocytes in the blood:

(a) comprise about a quarter of all the granulocytes
(b) account for most of the increase in white cell count caused by acute infections

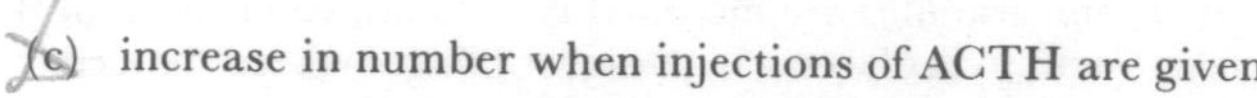

(c) increase in number when injections of ACTH are given

(d) are sites of antibody manufacture.

25. For blood clotting to occur normally:

(a) heparin must be inactivated

(b) there must be a sufficient dietary intake of vitamin C

(c) calcium ions must be present
(d) the liver must have an adequate supply of vitamin K.

26. Antibodies to the A and B antigens of red blood cells:

(a) are present in fetal plasma

(b) have a molecular weight similar to the average molecular weight of plasma globulins (150 000)
(c) pass easily from maternal to fetal blood across the placental barrier
(d) cause immediate lysis of A and B erythrocytes in the presence of complement.

27. Lymph:

(a) contains plasma proteins
(b) vessels are important in the absorption of protein from the intestine
(c) production in muscle is increased during muscular activity

(d) is cell-free.

28. Blood platelets:

(a) are formed in the bone marrow
(b) tend to increase in number after tissue damage due, e.g., to injury or a surgical operation
(c) have a small nucleus
(d) can alter their shape when they make contact with collagen.

24.
(a) F just 1 – 3% of all granulocytes
(b) F acute infections often raise the neutrophil granulocyte count; the eosinophil count may be raised in allergic conditions and in certain parasitic infestations
(c) F ACTH and cortisone injections lower the eosinophil count
(d) F eosinophils are thought to phagocytose products of antigen-antibody reactions.

25.
(a) F but the anticoagulant effects of heparin must be overwhelmed
(b) F the spontaneous bleeding that occurs from the gums etc. in scurvy is thought to be due to increased capillary fragility
(c) T removal of calcium ions prevents clotting
(d) T this is necessary for prothrombin formation.

26.
(a) F they are formed shortly after birth, possibly because the corresponding antigens enter the body in food or bacteria
(b) F the iso-agglutinins of the blood groups A and B are IgM antibodies (molecular weight 900 000)
(c) F otherwise serious fetal damage would be common, e.g. mother group O, fetus group A
(d) F they cause immediate agglutination (clumping) of A and B erythrocytes.

27.
(a) T it returns these to the blood
(b) F they are involved in the absorption of fat
(c) T dilatation of the resistance vessels increases capillary blood pressure and the formation of tissue fluid
(d) F it contains lymphocytes formed in the lymph nodes.

28.
(a) T from megakaryocytes
(b) T leading to an increased risk of intravascular blood clotting after injury or a surgical operation
(c) F they have no nucleus
(d) T they swell, throw out pseudopodia and adhere to the tissue and one another.

29. The granulocytes in the blood of normal people:
(a) number 250–500×10⁹/litre (250 000–500 000 per mm³)

(b) are capable of crossing capillary walls
(c) are multinucleated

(d) are derived from bone marrow.

30. Antigen-antibody reactions may result in:
(a) clumping of the cells carrying the antigen
(b) lysis of the cells carrying the antigen
(c) susceptibility of the cells carrying the antigen to phagocytosis
(d) precipitation of the antigen.

31. The life span of a human red cell:
(a) can be estimated from the survival time of Group A cells in a Group O recipient
(b) can be determined by giving a person a dose of radioactive iron and monitoring the time course of the ensuing radioactivity in red cells
(c) is, on average, 12 weeks
(d) is increased by erythropoietin.

32. Erythrocyte production:
(a) is stimulated when a person ascends to a height where the atmosphere pressure is halved

(b) can occur in the spleen

(c) is dependent on normal gastric secretory activity

(d) is stimulated by a rise in the pressure of carbon dioxide (P_{CO_2}) in the arterial blood.

29.
(a) F this is true of platelets; the average granulocyte count is around 3–6×10⁹/litre (3000–6000 per mm³)
(b) T this is necessary for their function
(c) F but the nucleus is multilobed, the number of lobes increasing with age
(d) T destruction of bone marrow is associated with a low granulocyte count.

30.
(a) T such antibodies are called *agglutinins*
(b) T such antibodies are called *lysins*
(c) T such antibodies are called *opsonins*
(d) T such antibodies are called *precipitins*.

31.
(a) F the cells would rapidly be haemolysed by the recipient's anti-A antibody
(b) T radioactivity rises as the dose is incorporated in a new generation of cells and falls when that generation is broken down in the reticuloendothelial system
(c) F it is about 17 weeks
(d) F erythropoietin regulates the rate of erythrocyte production, not the life span.

32.
(a) T the resulting oxygen lack leads to increased erythropoietin production and so to a raised red cell count in blood
(b) T splenic erythropoiesis occurs in the fetus; in normal adults, red cell formation is confined to bone marrow
(c) T gastric juice contains intrinsic factor which is necessary for the absorption of vitamin B_{12}
(d) F oxygen lack, not CO_2 excess, stimulates the production of erythropoietin. At altitude, erythropoiesis is stimulated though the P_{CO_2} is reduced.

33. The conversion of fibrinogen to fibrin:

(a) is promoted by prothrombin, a proteolytic enzyme

(b) involves the disruption of certain peptide linkages in fibrinogen

(c) is followed by polymerisation of the fibrin to form strands

(d) is inhibited by heparin, a strongly electronegative substance.

34. Plasma proteins:

(a) include certain clotting factors

(b) of the immunoglobulin variety are manufactured mainly in the liver

(c) include the protein carbonic anhydrase

(d) exert an osmotic pressure across the capillary wall nearer 1.0 than 20 milliosmoles/litre.

35. Blood:

(a) volume is 30–40 ml/kg in an adult of average build

(b) plasma contains more than 10 times as many sodium ions as protein molecules

(c) volume, as estimated by the dilution of an indicator such as radioactive albumin, appears greater if the dilution is measured 30 minutes rather than one hour after the indicator injection

(d) erythrocytes do not use oxygen for metabolism.

36. Thirst is:

(a) produced by a rise in plasma tonicity even though blood volume is normal

(b) produced by stimulation of certain areas in the hypothalamus

(c) produced by a fall in blood volume even though blood tonicity is normal

(d) relieved when water is drunk before the water has been absorbed and the blood volume and tonicity returned to normal.

33.

(a) F prothrombin has to be converted to thrombin before it can affect fibrinogen

(b) T thrombin breaks off the terminal groups which are linked to the rest of the protein fibrinogen by peptide bonds

(c) T this follows automatically when the fibrinogen has been changed to insoluble fibrin by loss of its terminal groups

(d) T heparin prevents the action of thrombin; electropositive substances (e.g. protamine) antagonise heparin's action.

34.

(a) T e.g. fibrinogen, antihaemophilic globulin and prothrombin

(b) F they are manufactured in the lymphatic system (the other plasma proteins are made in the liver)

(c) F this enzyme is found in the red cells

(d) T the colloid osmotic pressure (about 1.0 milliosmole/litre) is very small as compared with total plasma osmolality (about 280 milliosmoles/litre).

35.

(a) F it is around 75–80 ml/kg

(b) T more than 100 times as many

(c) F albumin steadily leaks from capillaries; hence its volume of distribution increases with time

(d) F like other living cells in the body they use oxygen, e.g. energy is required to pump ions across the red cell membrane.

36.

(a) T thirst is one consequence of stimulation of osmoreceptors

(b) T the osmoreceptors are situated in this region

(c) T in this situation, as with hypertonicity, there is need to increase water intake and reduce loss

(d) T this provides rapid negative feedback to the thirst-regulating region; of course the thirst returns if blood volume and tonicity remain abnormal.

37. An appropriate dilution indicator for measuring:

(a) total body water would be sucrose

(b) plasma volume would be radioactive sodium

(c) extracellular volume would be inulin

(d) total body potassium would be radioactive potassium.

38. Coagulation of blood *in vivo*:

(a) brought about solely by circulating factors is regarded as following the intrinsic rather than the extrinsic pathway

(b) can be initiated by the products of tissue damage

(c) leads to decreased formation of plasmin from plasminogen

(d) is followed by a fall in the circulating level of fibrin degradation products.

37.
(a) F sucrose would not cross the cell membrane freely for equilibration with intracellular fluid
(b) F the sodium ions would cross capillary walls freely into the interstitial fluid
(c) T inulin crosses capillary walls freely but is not taken up by cells (the same properties make it useful for measuring the glomerular filtration rate)
(d) T the body does not distinguish between isotopes of the same element; thus the radioactive isotope is eventually distributed throughout the total body potassium.

38.
(a) T this pathway includes factor VIII (antihaemophilic globulin)
(b) T this extrinsic pathway (system) bypasses the initial steps of the intrinsic system
(c) F formation of plasmin is increased
(d) F there is a rise, indicating the activity of plasmin in lysing the clot.

BODY FLUIDS
APPLIED PHYSIOLOGY

39. Intravenous infusion:
(a) of normal saline is the appropriate treatment for a patient who has just vomited 2 litres of blood from a bleeding gastric ulcer
(b) of bicarbonate is likely to be helpful in a patient who is being treated for acute cardiac and respiratory arrest
(c) fluids for a patient who has had prolonged vomiting should be potassium-free
(d) of a glucose solution isotonic with blood is appropriate treatment for a patient suffering from severe water depletion who can not take water by mouth.

40. Excessive tissue fluid (oedema) in the legs may result from:
(a) exposure of the legs to subatmospheric pressures
(b) a high arterial blood pressure in the absence of heart failure
(c) varicose veins in the legs
(d) blockage of the pelvic lymphatics.

41. Haemolytic disease of the newborn:
(a) is more common than average in infants who have the same ABO blood group as their mother
(b) is characterised by jaundice which improves spontaneously immediately after birth
(c) has been made rarer by treating appropriate mothers with anti-D antibody after delivery
(d) is treated by exchange transfusion with blood of the same ABO and Rhesus group as the infant.

42. The neutrophil granulocyte count rises:
(a) during exercise
(b) following cardiac infarction (death of an area of heart muscle due to loss of its blood supply)
(c) when the lymphocyte count rises
(d) with infections that lead to pus formation (suppuration).

39.
(a) F the saline will expand the depleted blood volume only temporarily; whole blood should be given
(b) T it corrects the acidosis due to lactic acid and CO_2 accumulation in the tissues
(c) F potassium lost in the vomitus (desquamated cells) and via the kidney (because of the alkalosis) should be replaced
(d) T since the solution is isotonic with blood it can be given intravenously with safety; when the glucose is metabolised the water remains.

40.
(a) T this lowers the tissue pressure without affecting the intravascular pressure so that the hydrostatic pressure gradient across the capillary wall is increased
(b) F in hypertension, arteriolar constriction results in raised arterial pressure; capillary pressure is normal
(c) T valvular incompetence causes venous, and hence capillary, pressure to rise in dependent limbs
(d) T protein accumulates in the interstitial fluid; the net colloid osmotic gradient across the capillary wall is reduced.

41.
(a) T ABO compatible cells are not destroyed by the mother and may cause Rhesus sensitisation
(b) F until birth the mother's liver excretes the bilirubin; jaundice deepens rapidly after birth
(c) T this prevents D-positive cells in the mother's circulation from sensitising her
(d) F Rhesus-positive blood would be attacked by the antibodies in the infant's circulation; Rhesus-negative blood is used.

42.
(a) T exercise is thought to mobilise granulocytes from stores
(b) T products of tissue damage cause the granulocyte count to rise
(c) F neutrophil and lymphocyte counts can vary independently
(d) T pus consists mainly of dead neutrophil granulocytes.

43. In the acid-base diagram below, where U and L represent the upper and lower levels of normal, someone whose arterial blood is represented by point:

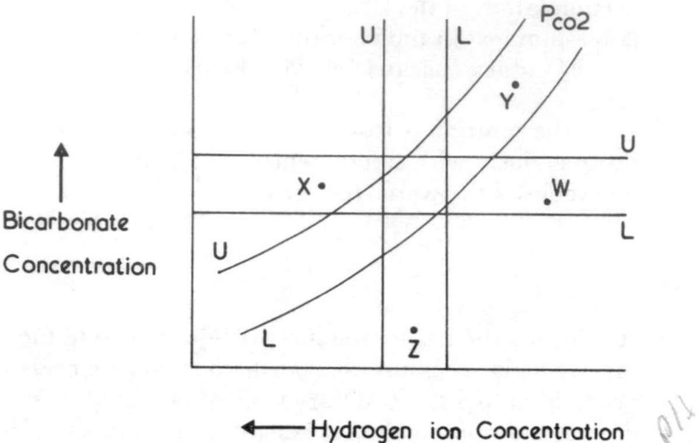

(a) W could be a normal person who has just completed a period of voluntary hyperventilation
(b) X could be a patient who has recently commenced excessive artificial ventilation
(c) Y could be a patient who has recently ingested a large dose of sodium bicarbonate
(d) Z could be suffering from either renal or chronic respiratory failure.

44. The appearance of blood which has sedimented for an hour may suggest:
(a) plasma albumin deficiency if the erythrocyte sedimentation rate is increased
(b) a high level of plasma lipids
(c) that the patient suffers from jaundice
(d) that haemolysis has occurred.

43.

(a) T there is an uncompensated respiratory alkalosis

(b) F the ventilation would be inadequate – there is an uncompensated respiratory acidosis

(c) T there is an uncompensated metabolic (or more strictly non-respiratory) alkalosis

(d) F the first is a possibility (compensated metabolic acidosis) but not the second (rather there is the possibility of a compensated respiratory *alkalosis*).

44.

(a) F an increased ESR suggests an abnormal pattern of plasma globulins

(b) T in which case the plasma appears cloudy or even milky

(c) T the plasma will be yellow

(d) T the plasma will be red; this may be due to a faulty technique in sampling the blood, e.g. having some water in the syringe.

45. Patients with anaemia tend to have:

(a) a low cardiac output
(b) an increased incidence of heart murmurs and bruits (sounds heard over large arteries)
(c) pallor of the mucous membranes

(d) an arterial Po_2 which is reduced in proportion to the reduction in haemoglobin level.

46. Appropriate treatment for a patient whose blood has an increased tendency to clot includes:

(a) intravenous heparin

(b) intravenous sodium citrate

(c) administration of a vitamin K antagonist

(d) strict bed rest.

47. Iron deficiency anaemia:

(a) is commoner in men than in women

(b) is characterised by large, pale erythrocytes
(c) should generally be treated by intramuscular injections of iron
(d) is the form of anaemia typically found following chronic blood loss from the body.

48. Transfusion of blood:

(a) of group O Rhesus negative to a patient of group AB Rhesus positive generally causes a severe reaction
(b) which is Rhesus positive to a Rhesus-negative girl of ten who has had no earlier transfusions is likely to cause a reaction
(c) which gives rise to a severe transfusion reaction is likely to be followed by jaundice
(d) to a patient who has had frequent previous transfusions requires special care in cross-matching.

45.
(a) F there is a compensatory increase
(b) T the increased velocity of blood flow and decreased viscosity lead to increased turbulence
(c) T due to a reduced haemoglobin concentration in the superficial vessels
(d) F Po_2 is little affected, *content* is reduced in proportion to the reduction in haemoglobin level.

46.
(a) T this antagonises thrombin and is effective within a few minutes
(b) F this would produce death from calcium ion deficiency before it would interfere with clotting
(c) T this leads to prothrombin deficiency. (These drugs are usually given by mouth)
(d) F this favours stasis of blood in the veins and clotting.

47.
(a) F it is commoner in women on account of the menstrual loss
(b) F they are small and pale
(c) F oral iron is avidly absorbed and is generally adequate
(d) T iron is the most critical constituent of the blood so lost.

48.
(a) F it is reasonably safe and is done in emergencies
(b) F but it may sensitise her and cause her children to have haemolytic disease if she marries a Rhesus-positive man
(c) T due to red-cell breakdown
(d) T because he may be sensitised to a number of blood group antigens other than those of the ABO and Rhesus systems.

49. Red cells:

(a) from healthy individuals are lysed when put in a solution of urea which is isosmolar with plasma

(b) from patients with sickle cell anaemia contain haemoglobin with an abnormal amino acid sequence

(c) are smaller than normal in patients with folic acid deficiency

(d) are normal in size in patients with anaemia due to depression of bone marrow activity.

50. Haemopoietic activity in the bone marrow is typically increased:

(a) following injections of vitamin B_{12} into a healthy individual on a normal diet

(b) a week after a large haemorrhage

(c) if the reticulocyte count is high in the peripheral blood

(d) in a patient suffering from haemolytic anaemia.

51. Vitamin B_{12} deficiency leads to:

(a) anaemia in which the red cells are smaller and paler than normal

(b) wasting of the gastric mucosa so that production of both hydrochloric acid and intrinsic factor is deficient

(c) a reduction in the circulating level of neutrophils and platelets

(d) loss of myelin from nerves.

52. In terms of the ABO blood groups, the transfused cells are likely to be affected by an antigen-antibody reaction when:

(a) group A blood is transfused to a group B person

(b) group O blood is transfused to a group AB person

(c) group A blood is transfused to a group O person

(d) group A blood is transfused to a group AB person.

53. The lymphocyte count:

(a) is affected less by irradiation than is the granulocyte count

(b) falls when immunosuppressive drugs are administered

(c) falls abruptly following removal of the thymus gland in the adult

(d) is low in animals who had their thymus glands removed in infancy.

49.
(a) T urea readily penetrates the red cell membrane; the osmotic force due to the red cell electrolytes draws water in and causes lysis
(b) T this haemoglobin S precipitates at low oxygen tensions so that the cells change shape and are readily haemolysed
(c) F folic acid deficiency causes a macrocytic anaemia
(d) T the cells are normocytic and normochromic.

50.
(a) F in such an individual B_{12} is not limiting haemopoietic activity
(b) T to correct the red cell deficit
(c) T a high reticulocyte count indicates an active marrow
(d) T to replace the haemolysed cells.

51.
(a) F the cells are large and are not pale
(b) F this is the *cause* not the consequence of B_{12} deficiency in "pernicious anaemia"; there are other causes of B_{12} deficiency
(c) T B_{12} is used in DNA synthesis which is demanded at a high rate by all rapidly multiplying cells
(d) T this leads to malfunction in the nervous system, e.g. subacute combined degeneration of the spinal cord.

52.
(a) T the recipient will have anti-A antibody
(b) F group O people are "universal donors"
(c) T a group O person will have anti-A antibodies which will agglutinate the transfused blood
(d) F group AB persons are "universal recipients", having no ABO antibodies.

53.
(a) F the lymphocyte count is very sensitive to irradiation
(b) T lymphocytes and immune responses are closely related
(c) F there is little active thymic tissue in the adult
(d) T it is thought that the thymus "seeds" the other lymphoid tissues with lymphocytes.

54. In the acid-base diagram below, where U and L represent the upper and lower levels of normal respectively, someone whose arterial blood is represented by point:

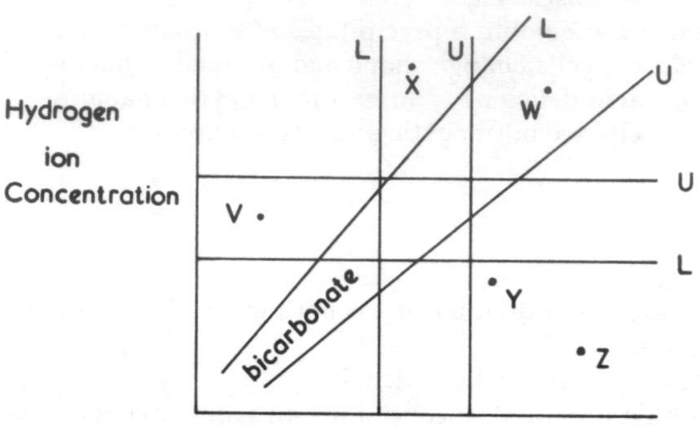

(a) V might have a compensated metabolic alkalosis

(b) W might have a compensated respiratory acidosis

(c) X shows no evidence of normal respiratory compensation

(d) Y shows more compensation than someone whose arterial blood is represented by point Z.

55. If one litre of group B blood is transfused into a group O recipient:

(a) aggregates of agglutinated cells may block a pulmonary artery

(b) the recipient is likely to develop fever

(c) urinary output will be diminished or suppressed

(d) the mistake is unlikely to endanger life.

54.

(a) F the low bicarbonate level indicates a metabolic *acidosis*, the carbon dioxide partial pressure is low in compensation

(b) F the respiratory acidosis is uncompensated since the hydrogen ion concentration is raised and the bicarbonate level normal

(c) T there is a severe metabolic acidosis but no lowering of the carbon dioxide level

(d) F Z could represent a severe partly compensated metabolic alkalosis/respiratory acidosis; Y shows a corresponding but much milder departure from normal.

55.

(a) F the cells do not clump to form large emboli but undergo lysis

(b) T due to release of fever-producing substances (pyrogens) from damaged cells

(c) T due to renal damage which is usually associated with haemoglobinuria

(d) F a rapidly fatal outcome is likely.

56. In a patient with a red cell count (RCC) of 4×10^{12}/litre, a haemoglobin (Hb) of 7.5 g/100 ml and a haematocrit of 0.28:

(a) the mean corpuscular haemoglobin (MCH) is nearer 20 picograms (pg) than 20 nanograms (ng). (1 pg $= 10^{-12}$ g; 1 ng $= 10^{-9}$ g)

(b) the mean cell volume (MCV) is nearer 95 than 70 fl (1 fl, femtolitre, = 1 μm^3)

(c) the mean corpuscular haemoglobin concentration (MCHC) is nearer 30 than 35 g/100 ml

(d) the cause of the anaemia is most likely to be vitamin B_{12} deficiency.

57. The tendency for a transplanted organ or tissue to be rejected is reduced:

(a) by treatment which causes a low blood lymphocyte count

(b) by keeping the recipient in isolation in a germ-free environment

(c) by irradiation of the organ or tissue with X-rays

(d) by drugs which interfere with mitosis.

58. The neutrophil granulocyte count tends to fall:

(a) when drugs are used which depress the bone marrow

(b) in pernicious anaemia (vitamin B_{12} deficiency)

(c) in response to trauma

(d) in response to bacterial infection.

56.

(a) T $MCH = \frac{Hb/100\ ml}{RCC/100\ ml} = \frac{7.5}{4\times10^{11}\ g}$

$= 18.75\times10^{-12}\ g = 18.78\ pg$

This is below normal (27–32 pg)

(b) F $MCV = \frac{\text{red cell volume/litre}}{\text{red cell count/litre}} = \frac{0.28}{4\times10^{12}}\ \text{litres} = 70\ fl$

since the normal volume is about 75–95 fl, these cells are microcytic

(c) T $MCHC = \frac{Hb}{\text{Haematocrit}} = \frac{7.5}{0.28} = 26.8\ g/100\ ml;$

since the normal MCHC is about 30–35 g/100 ml, these cells are low in haemoglobin

(d) F this microcytic hypochromic picture is characteristic of iron deficiency.

57.

(a) T the lymphocyte series is responsible for the immunological response to foreign proteins

(b) F patients are kept in a germ-free environment because immunosuppression reduces their resistance to infection

(c) F this could damage the transplant without reducing its antigenic effect

(d) T these suppress lymphocyte multiplication.

58.

(a) T bone marrow is the sole source of neutrophil granulocytes

(b) T vitamin B_{12} is needed for all cell series that multiply by rapid mitosis

(c) F products of tissue destruction usually stimulate neutrophil production

(d) F a high granulocyte count is usually seen with bacterial infection.

59. In jaundice due to excessive haemolysis of erythrocytes there is:
(a) a decrease in the amount of conjugated bilirubin excreted by the liver
(b) an increased amount of stercobilin in the faeces
(c) a decreased amount of urobilinogen in the urine
(d) an increase in the protein-bound bilirubin in the blood.

60. A fall in plasma sodium concentration:
(a) tends to lower the freezing point of plasma
(b) tends to cause intracellular fluid volume to increase
(c) is associated with excessive thirst
(d) can be caused by excessive (inappropriate) secretion of antidiuretic hormone (ADH).

61. Depletion of body potassium:
(a) is more accurately detected by analysis of a muscle biopsy than by measurement of the plasma potassium level
(b) tends to cause increased activity of intestinal smooth muscle
(c) tends to cause acidosis
(d) is associated with decreased amplitude of the T wave of the electrocardiogram.

62. A high blood potassium level (hyperkalaemia):
(a) can be reduced by the administration of glucose and insulin
(b) is a typical consequence of persistent severe diarrhoea
(c) tends to occur when a person has been trapped in a situation where limbs are crushed
(d) is commonly seen in acute renal failure even though the patient is given a potassium-free diet.

59.
(a) F it is increased due to the increased bilirubin load presented to the liver
(b) T due to the increased excretion of bilirubin into the gut
(c) F it is increased due to increased reabsorption of stercobilinogen from the gut
(d) T bilirubin released by red cell breakdown is made soluble by binding to plasma albumin.

60.
(a) F freezing point rises towards 0°C as plasma osmolality (highly correlated with sodium concentration) falls
(b) T the extracellular fluid tends to be hypotonic; water is then drawn into cells osmotically and may result in water intoxication
(c) F hypertonicity of extracellular fluid is associated with thirst
(d) T ADH suppresses water diuresis so that excessive water is retained and dilutes both extra- and intracellular fluid.

61.
(a) T the great bulk of body potassium is intracellular; plasma levels may not parallel intracellular levels. The biopsy method is not, of course, practical for routine use
(b) F activity tends to decrease; in severe cases, intestinal muscle paralysis (paralytic ileus) may result
(c) F since K^+ competes with H^+ for secretion by the renal tubules, hypokalaemia leads to excessive loss of H^+ in the urine, hence, alkalosis
(d) T this together with depression of the ST segment is a useful index of hypokalaemia.

62.
(a) T insulin facilitates the entry into cells of glucose and potassium
(b) F severe diarrhoea causes K^+ depletion since there is excessive loss of the K^+ in intestinal fluids and cellular debris
(c) T large amounts of potassium are released from the damaged muscles
(d) T normal cellular wear and tear releases K^+ which accumulates in the blood since it cannot be excreted in the urine.

63. A plasma bicarbonate level 50% of normal would be consistent with a diagnosis of:
(a) compensated respiratory acidosis
(b) metabolic acidosis
(c) compensated metabolic acidosis
(d) compensated respiratory alkalosis.

64. Carbon monoxide:
(a) has nearer 200 than 20 times the affinity of oxygen for haemoglobin
(b) shifts the oxygen dissociation curve of haemoglobin to the right
(c) in the inspired air acts as a stimulus to ventilation
(d) in cigarette smoke can result in 5–10% of the smoker's haemoglobin being bound to carbon monoxide.

65. A severe reduction in the number of circulating neutrophil polymorphonuclear granulocytes (neutrophils) is:
(a) present if the total white cell count is 60 000/mm^3 and the differential white cell count shows 10% neutrophils
(b) associated with a sore throat
(c) associated with purulent throat, mouth and skin infections
(d) caused by a high level of circulating glucocorticoids.

66. A reduction in the blood level of coagulation factor VIII (antihaemophilic globulin):
(a) typically increases the bleeding time beyond the normal range
(b) occurs as a hereditary disease (haemophilia) due to an abnormal gene on the Y chromosome
(c) to 80% of average would account for excessive bleeding after tooth extraction
(d) is typically associated with small petechial haemorrhages into the skin (purpura).

63.

(a) F in this condition the bicarbonate level is raised

(b) T a low bicarbonate level is the hallmark of metabolic acidosis

(c) T in compensated metabolic acidosis, the $[HCO_3]$ is still reduced; the compensation which restores the pH to near normal is a fall in the P_{CO_2}

(d) T the pH change caused by the fall in P_{CO_2} in respiratory alkalosis is compensated for by an increased renal excretion of bicarbonate.

64.

(a) T the ratio is between 200 and 300

(b) F it shifts it to the left so that less oxygen is released to the tissues at a given oxygen pressure

(c) F there is no such stimulation of ventilation

(d) T exposure to heavy exhaust gas pollution in traffic can have the same effect.

65.

(a) F these figures indicate 6000 neutrophils/mm^3, near the upper limit of normal

(b) T the bacteria which are constantly being deposited in the pharynx are not eliminated as efficiently as usual

(c) F the infections are notable for an absence of pus (dead neutrophils which have ingested bacteria)

(d) F glucocorticoids tend to cause a marked reduction in lymphocyte and eosinophil counts.

66.

(a) F bleeding time is determined by platelet function and vascular contraction rather than blood coagulation

(b) F the pattern of heredity is consistent with a recessive abnormality of the X chromosome

(c) F bleeding manifestations commence only when the level has fallen below 50% of average normal

(d) F purpura is typical of haemorrhagic disorders due to capillary and platelet abnormalities.

67. Sodium ion retention is associated with:

(a) an increase in the percentage of the total body water that is in the extracellular compartment
(b) a rise in capillary blood pressure
(c) mineralocorticoid but not glucocorticoid administration

(d) the first 3–5 days after a major surgical operation.

68. Acidosis:

(a) tends to occur when the ratio $|H_2CO_3|:|HCO_3^-|$ falls below normal
(b) of the compensated metabolic variety is associated with low levels of both HCO_3^- and P_{CO_2}
(c) of the compensated respiratory variety is associated with a raised level of HCO_3^-
(d) increases the risk of a low level of ionised calcium and hence tetany.

69. In the acid-base diagram below where L and U represent the lower and upper levels of normal respectively, a person whose arterial blood values were found to be at point:

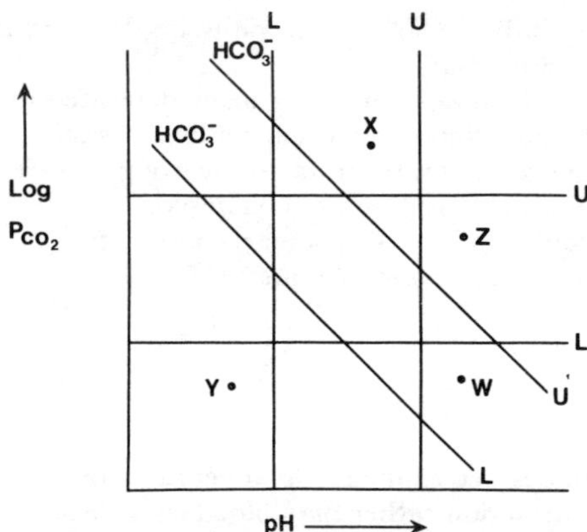

(a) W would have uncompensated respiratory alkalosis

(b) X might have a compensated metabolic alkalosis
(c) Y might have a partly compensated respiratory acidosis
(d) Z might be suffering from severe vomiting.

67.
(a) T sodium is mainly extracellular, retention of sodium is accompanied by retention of chloride and water
(b) T this causes oedema
(c) F both groups cause sodium retention; glucocorticoids also have mineralocorticoid effects
(d) T this is part of the metabolic response to trauma.

68.
(a) F this change tends to cause an alkalosis
(b) T the low $|HCO_3^-|$ is primary, the fall in P_{CO_2} the secondary, compensating factor
(c) T compensation in this case is due to a rise in $|HCO_3^-|$
(d) F acidosis decreases the binding of calcium to the plasma proteins and so raises the level of ionic calcium.

69.
(a) T the rise in pH is associated with a low P_{CO_2} but a normal $|HCO_3^-|$
(b) T or a compensated respiratory acidosis
(c) F he has a partly compensated metabolic acidosis
(d) T he has an uncompensated metabolic alkalosis.

70. In investigating an acid-base disturbance:

(a) observations should be made on venous rather than arterial blood

(b) blood samples should be stored in ice until it is possible to make the appropriate measurements

(c) the pH can be calculated if the P_{CO_2} and the $|HCO_3^-|$ are estimated

(d) a raised level of ammonium salts in the urine would be consistent with the diagnosis of a metabolic acidosis.

71. Acidosis may result in:

(a) potassium retention

(b) a rise in the plasma chloride level

(c) a low P_{CO_2}

(d) peripheral vasodilatation.

72. A high level of ionised calcium in the blood:

(a) is associated with increased excitability of muscle and nerve

(b) may result from chronic renal failure

(c) may result in the deposition of stones in the urinary tract

(d) is seen following injections of gland extracts from the anterior pituitary.

70.

(a) F acid-base balance is regulated with precision only in arterial blood. The acid-base status of venous blood is determined by the tissue which it drains

(b) T this reduces the acid-base changes produced by continuing red cell metabolism

(c) T the variables are related: $pH \propto \frac{|HCO_3^-|}{P_{CO_2}}$

(d) T the kidney secretes ammonia to buffer the increased number of hydrogen ions being excreted in metabolic acidosis.

71.

(a) T because the raised $|H^+|$ competes with K^+ for secretion by the renal tubules

(b) T in metabolic acidosis, the bicarbonate anion deficit in the plasma is made up by chloride

(c) T in metabolic acidosis, the compensatory hyperventilation reduces the P_{CO_2} so that the $|H_2CO_3| : |HCO_3^-|$ ratio tends to return towards 1:20

(d) T this is seen in respiratory acidosis due to CO_2 retention. To maintain arterial blood pressure there is an increase in cardiac output which may contribute to the heart failure in cor pulmonale.

72.

(a) F the reverse is true

(b) F the phosphate retention in chronic renal failure tends to lower the $|Ca^{2+}|$ in blood ($|Ca^{2+}|\ |PO_4^{3-}| = K$)

(c) T when Ca^{2+} and PO_4^{3-} exceed their solubility product calcium phosphate may be precipitated in the kidney and other tissues

(d) F the pituitary gland does not regulate calcium levels in blood.

73. Sodium depletion differs from sodium retention in that:

(a) central venous pressure (CVP) tends to be low

(b) the renal production of renin tends to be high

(c) the specific gravity of the blood is reduced

(d) the intracellular fluid (ICF) volume tends to decrease.

74. Respiratory alkalosis differs from metabolic alkalosis in that:

(a) there is less danger of tetany
(b) the urine is alkaline in reaction
(c) the plasma bicarbonate level is normal or low

(d) there is a greater tendency for cerebral blood flow to fall.

75. The effects of sodium depletion differ from those of water depletion in that:

(a) cardiovascular changes are less pronounced

(b) intracellular fluid is less depleted

(c) the haematocrit tends to be higher
(d) thirst tends to be greater.

76. A raised blood potassium level (hyperkalaemia) differs from a low blood potassium level (hypokalaemia) in that it causes:

(a) skeletal muscular weakness

(b) a tendency to alkalosis

(c) mental confusion and apathy
(d) myocardial weakness leading possibly to heart failure.

73.
(a) T sodium loss is associated with water loss and a fall in the extracellular fluid (ECF) volume. The resulting fall in plasma volume reduces the CVP and may result in peripheral circulatory failure
(b) T the fall in blood pressure is detected by the juxtaglomerular apparatus in the kidney
(c) F the decrease in plasma volume increases the haematocrit since the cell volume is not affected
(d) F the tendency for the tonicity of the ECF to fall results in the osmotic movement of fluid *into* the cells.

74.
(a) F both types of alkalosis may cause tetany
(b) F it is alkaline in both
(c) T it is initially normal; a compensatory increase in the renal excretion of bicarbonate tends to bring the $|H_2CO_3|:|HCO_3^-|$ ratio back to 1:20
(d) T due to the loss of the cerebral vasodilator effects of carbon dioxide.

75.
(a) F the blood volume tends to be reduced to a greater extent in sodium depletion
(b) T the hypertonicity of the ECF in water depletion draws water from the ICF
(c) T due to the more severe reduction in plasma volume
(d) F the main stimulus to thirst is hypertonicity of the ECF.

76.
(a) F both hyper- and hypokalaemia cause skeletal muscular weakness
(b) F hypokalaemia causes alkalosis since potassium competes with hydrogen ion for excretion in the renal tubules
(c) F both cause depression of the mental state
(d) F both can lead to myocardial weakness, abnormal rhythms and heart failure.

77. Intravenous infusion of one litre of:

(a) 5% dextrose provides the patient with nearer 4.2 than 0.42 megajoules (nearer 1000 than 100 calories)
(b) a lipid suspension provides between two and three times as much energy as the same concentration of carbohydrate
(c) an amino acid solution provides between three and four times as much energy as the same concentration of carbohydrate

(d) 10% dextrose would provide the daily energy requirement of the average adult.

78. Release into the circulation of large quantities of cellular breakdown products (due, e.g. to severe tissue damage or a "concealed" placental haemorrhage) tends to cause:

(a) coagulation of blood via the "extrinsic" pathway

(b) widespread laying down of the products of blood coagulation
(c) a raised level of circulating fibrinogen

(d) a haemorrhagic state.

77.

(a) F the 50 grams of dextrose provide approximately 0.84 megajoules (200 calories).

(b) T just over twice as much

(c) F there is little difference in the energy provided by the two solutions; amino acids are given to maintain body protein, not to provide energy

(d) F this would provide only about one fifth of the energy requirement.

78.

(a) T tissue products, rather than circulating factors initiate coagulation

(b) T hence the term "disseminated intravascular coagulation"

(c) F the fibrinogen laid down as fibrin is removed from the circulation

(d) T because the coagulation factors, including fibrinogen, have been used up, hence the other term "consumptive coagulopathy".

CARDIOVASCULAR SYSTEM BASIC PHYSIOLOGY

79. Blood flow through the left coronary artery:
(a) is greater in systole than diastole

(b) is regulated by sympathetic vasodilator nerves
(c) increases in the presence of myocardial hypoxia

(d) is decreased in the reflex vasoconstrictor response to a fall in systemic arterial blood pressure.

80. In a healthy individual the metabolic requirements of the tissues may exceed the capacity of the circulation to deliver blood in:
(a) the heart

(b) the skin

(c) the muscles

(d) the kidneys.

81. The pressure:
(a) drop across the major veins is similar to that across the major arteries
(b) drop across the hepatic portal bed is similar to that across the splenic vascular bed
(c) in the hepatic portal vein is higher than that in the inferior vena cava
(d) drop across the vascular bed in the foot is greater when standing up than when lying down.

82. The second heart sound differs from the first heart sound in that:
(a) it is due partly to turbulence set up by valve closure

(b) its duration is greater than that of the first sound
(c) it has a higher frequency

(d) it is occasionally split, i.e. the sound from the right side of the heart does not occur synchronously with that from the left.

79.
(a) F the reverse is true due to compression of vessels by the contracting left ventricle
(b) F these have not been demonstrated
(c) T in this way coronary blood flow is adjusted to meet the metabolic requirements of the myocardium
(d) F coronary, like cerebral, blood vessels are little affected in the vasoconstrictor response to arterial hypotension.

80.
(a) F even in severe exercise, coronary blood flow does not limit cardiac performance
(b) F the metabolic requirements of skin are very small, the circulation can deliver very high rates of flow to the skin in thermoregulatory reflexes
(c) T during severe exercise the skeletal muscles build up a severe oxygen debt
(d) F the normal flow to the kidneys (over 1 l/min) is far in excess of the most demanding renal metabolic needs.

81.
(a) T since they offer approximately similar resistance to flow
(b) F the drop across the hepatic portal system is much smaller
(c) T otherwise blood would not flow through the portal bed
(d) F when a subject stands up both the arterial and venous pressures rise by the hydrostatic equivalent of the column of blood below the heart. The pressure drop therefore stays the same.

82.
(a) F turbulence due to valve closure contributes to both heart sounds
(b) F its duration is about 20% less than that of the first sound
(c) T about 50 Hz compared with about 35 Hz for the first sound
(d) F both may be split in health due to asynchronous valve closure.

83. The panels below show records of two measurements of cardiac output by the indicator dilution method. Injection of 10 mg of indicator into the right atrium took place at time 0. The concentration of the indicator in arterial blood (mg/litre) has been plotted at intervals after injection. The records show:

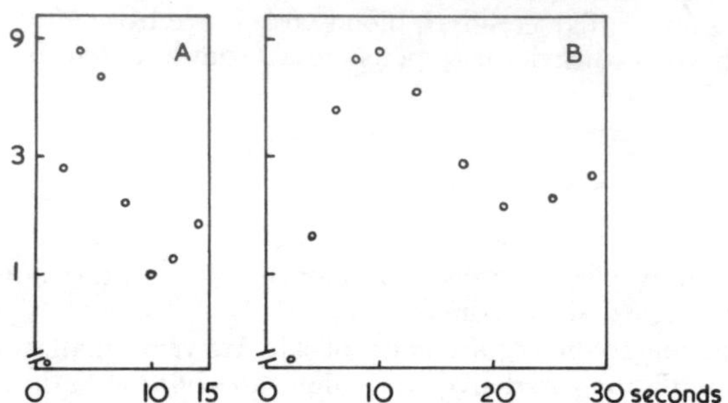

(a) a concentration scale which is logarithmic

(b) a greater cardiac output in panel B than in panel A (assuming a similar injection in each case)

(c) evidence, in both cases, that some of the indicator had passed the sampling point twice

(d) a cardiac output in panel A which is nearer 5 than 10 litres per minute.

84. Pulmonary vascular resistance:

(a) is less than one-third that offered by the systemic circuit

(b) is decreased locally by a local fall in alveolar Po_2

(c) plays an important part in determining total pulmonary blood flow

(d) decreases during exercise.

83.

(a) T equal intervals on the scale represent a threefold increase in concentration

(b) F output is greater in A; the indicator passed the sampling point in a shorter time

(c) T recirculation is indicated by the rising concentration towards the ends of the records

(d) F since 10 mg indicator was injected, and mean concentration was 3–4 mg/litre (log scale!) the indicator was contained in about 3 litres of blood. This passed the sampling point in 12 seconds, giving a cardiac output of approximately 15 litres per minute.

84.

(a) T mean pulmonary perfusion pressure (ΔP) is less than one-quarter of the systemic value; Flow (F) is the same. Resistance $= \Delta P/F$

(b) F the reverse is true; this tends to divert blood to areas of the lungs where the alveolar P_{O_2} is higher

(c) F pulmonary blood flow (cardiac output) is determined by the requirements of the systemic circuit. Over a length of time it must equal left ventricular output

(d) T the increase in pulmonary blood flow in exercise is accommodated with little rise in pulmonary artery pressure, probably due to passive distension of the pulmonary resistance vessels.

85. Ventricular filling:

(a) depends mainly on contraction of the atria

(b) begins during isometric ventricular relaxation

(c) gives rise to a third heart sound in some healthy people

(d) will not occur unless atrial pressure is higher than atmospheric pressure.

86. The veins:

(a) receive their nutrition via vasa vasorum arising from the veins

(b) are innervated by sympathetic constrictor fibres

(c) have smooth muscle in their walls which is excited to contract by venous distension

(d) contain the major part of the blood volume.

87. In the heart:

(a) excitation cannot spread directly from atrial to ventricular muscle fibres

(b) there is appreciably more muscle in the left than in the right atrium

(c) ventricular muscle is arranged so that the contracting ventricle shortens from base to apex as well as reducing its diameter

(d) contraction normally begins in the right atrium.

88. In exercise using isometric as opposed to isotonic muscular contraction:

(a) the "muscle pump" action on venous return is more effective

(b) venous pressure in the active muscles tends to be lower

(c) there is a greater increase in mean arterial pressure

(d) there is a greater increase in cardiac work for the same increase in cardiac output.

85.
(a) F atrial contraction accounts for only about 20% of ventricular filling
(b) F during this phase the A-V valves are closed
(c) T the rapid filling phase in early diastole may give rise to a low-pitched sound in young people
(d) F ventricular filling can occur with subatmospheric atrial pressure provided that it is higher than ventricular pressure.

86.
(a) F they receive their nutrition from vasa vasorum arising from neighbouring arteries
(b) T these regulate venous capacity
(c) T this *myogenic* response helps to limit venous distension when intravenous pressure rises
(d) T around 75%.

87.
(a) T atrial and ventricular fibres are attached to a fibrous skeleton at the base of the heart; excitation spreads from atrial to ventricular fibres via a specialised band of tissue that passes through the fibrous skeleton, the bundle of His
(b) F the work load of the two atria is similar in health
(c) T the ventricular muscle is arranged as a sandwich of circular muscle between two coats of spiral muscle at right angles to one another
(d) T this is where the sino-atrial node is situated.

88.
(a) F the muscle pump is more effective when there is intermittent muscle contraction as in isotonic contraction (e.g. while running)
(b) F it is lower in isotonic contraction (dynamic exercise) because the muscle pump tends to empty the veins
(c) T this would help to drive blood through the vessels compressed by the isometrically contracted muscles in static effort
(d) T cardiac work for a given output is a function of mean arterial pressure.

89. The figure below shows simultaneous records of changes in right hand and left forearm volume in relation to the collecting pressure applied by means of cuffs on the right wrist and left upper arm respectively. The volume of hand in the plethysmograph was 300 ml and the volume of forearm was 600 ml. These records (venous occlusion plethysmograms) show:

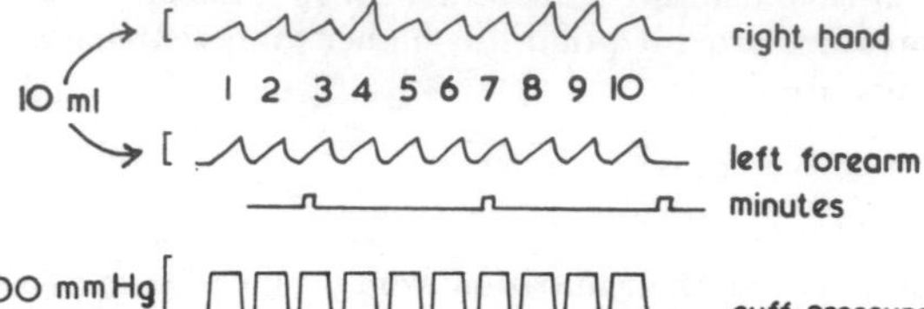

(a) a greater hand blood flow at point 3 than at point 4.

(b) that blood flow rate per unit volume in the hand is similar to that in the forearm

(c) more evidence for fluctuating sympathetic tone in the right hand record than in the left forearm record

(d) forearm blood flow rates nearer 20 than 50 ml/min.

90. The loss of fluid from capillary blood vessels in the legs is:

(a) increased when the arterioles in the legs constrict
(b) decreased when the subject changes from the recumbent to the standing position

(c) greater than the gain of fluid by these capillaries
(d) increased during leg exercise.

91. In the measurement of blood pressure by the auscultatory method:

(a) the first sounds heard as cuff pressure is lowered indicate the systolic pressure

(b) it is not possible to estimate diastolic pressure

(c) the values obtained for systolic pressure tend to be lower than those obtained using the palpatory method
(d) a cuff which is wider than normal may be required for very obese arms.

89.

(a) F flow is measured as increase in volume with time and hence is directly proportional to the slope of the volume record during collection

(b) F the total blood flow rates are similar on the two sides; since hand volume is half that of the forearm, its rate of flow per unit volume is approximately double.

(c) T this accounts for the rapid fluctuations of hand blood flow which are usually found with the subject resting in a comfortable environment

(d) F the rate of flow is quite close to 50 ml/min.

90.

(a) F this reduces capillary hydrostatic pressure

(b) F in the standing position the pressure in the leg capillaries increases by the hydrostatic equivalent of the column of blood below the heart

(c) T some of the fluid lost is removed by lymphatics

(d) T due to the rise in capillary pressure with arteriolar dilatation.

91.

(a) T the sharp taps of phase 1 are due to vibrations caused by the systolic pressure peaks forcing blood through the arteries under the cuff

(b) F diastolic pressure can be estimated from the Kortotkoff sounds and is thought to lie between sudden muffling (phase 4) and complete disappearance (phase 5)

(c) F they tend to be higher since palpation may fail to detect the first tiny pulses

(d) T otherwise the full cuff pressure may not be transmitted to the arteries; falsely high readings would be obtained.

92. The absolute refractory period in the heart:

(a) is the phase of the cardiac cycle when the heart cannot be stimulated by any stimulus, however strong
(b) corresponds approximately in time with the duration of the action potential
(c) is longer than the absolute refractory period in skeletal muscle

(d) lasts approximately as long as the cardiac contraction.

93. The resistance (R) offered by a blood vessel to blood flow:

(a) falls to one-eighth of its former value when the radius (r) of the vessel is doubled
(b) depends upon the thickness (T) of the wall of the vessel
(c) rises as the viscosity (η) of the blood rises
(d) is directly proportional to its length (L).

94. Activity in sympathetic nerves to the heart increases:

(a) reflexly whenever activity in parasympathetic nerves to the heart decreases
(b) during exercise

(c) during excitement

(d) when the arterial pressure falls.

95. Venous pressure in the:

(a) venous sinuses of the skull is subatmospheric when the head is above heart level

(b) foot of an adult who is standing still is approximately equal to his arterial pressure at heart level
(c) foot tends to be lower when the subject is walking than when he is standing still
(d) thorax decreases when the subject inhales deeply.

96. The increase in blood flow to a rhythmically exercising muscle depends on:

(a) release of sympathetic vasoconstrictor tone

(b) active dilatation of the capillaries

(c) vasoconstriction in vascular beds adjacent to the exercising muscle
(d) metabolites produced in the exercising muscle.

92.
(a) T this is its definition
(b) T the cells cannot be excited until their membranes have been repolarised
(c) T the action potentials and therefore the refractory periods are much longer in heart muscle than in skeletal muscle
(d) T this prevents tetanic contraction of the heart.

93.
(a) F it falls to $^1/_{16}$ $R \propto 1/r^4$
(b) F resistance is not a function of the thickness of the wall
(c) T $R \propto \eta$
(d) T to sum up $R \propto L\eta/r^4$ (from Poiseuille's law).

94.
(a) F sympathetic and parasympathetic nerves can function independently
(b) T beta adrenoceptor blockade reduces the increase in heart rate in exercise
(c) T beta adrenoceptor blockade also reduces the increase in heart rate with excitement
(d) T this is part of the reflex pressor response to diminished stimulation of the arterial stretch receptors.

95.
(a) T care has to be taken in brain operations lest damage to sinus walls allows air to be sucked into them and cause air embolism
(b) T due to the pressure exerted by the column of blood extending from the heart to the foot
(c) T due to the muscle pump
(d) T the fall in intrathoracic pressure is transmitted to the veins and aids venous return.

96.
(a) F exercise hyperaemia is normal in a sympathectomised muscle
(b) F the capillaries lack a muscle coat and there is no evidence that they can dilate actively
(c) F the increased flow is due to an increase in, rather than a redistribution of, cardiac output
(d) T these decrease vascular resistance by dilating arterioles.

97. The sino-atrial node:

(a) but not the atrioventricular node receives a parasympathetic nerve supply
(b) is connected by thin bands of Purkinje tissue with the A-V node
(c) pacemaker cells undergo a faster rate of diastolic depolarisation at 39°C than at 37°C
(d) pacemaker cells undergo a faster rate of diastolic depolarisation in the presence of acetylcholine.

98. The diagram below shows records of pressure in the right atrium, right ventricle and pulmonary artery of a normal heart. In this diagram:

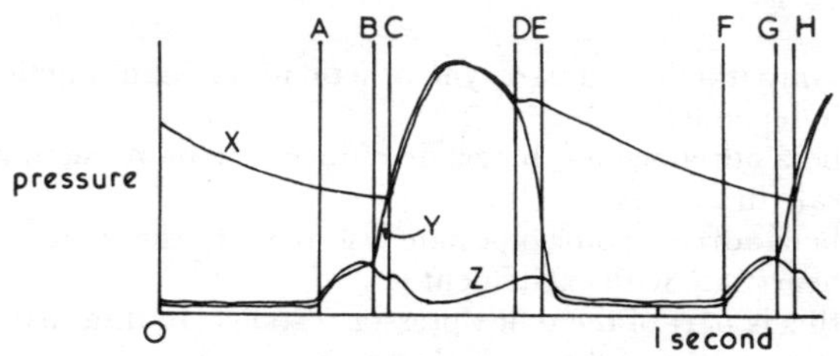

(a) line Y rather than line X represents right ventricular pressure
(b) heart rate in the cycle shown is above rather than below 60/minute
(c) the pressure peak of line Y would correspond more closely to 20 than to 50 mmHg
(d) pulmonary valve opening occurs at points B and G.

99. In the above diagram:

(a) pulmonary valve closure occurs at point E
(b) isometric relaxation time is longer than isometric contraction time
(c) the tricuspid valve is open at points A and F
(d) the pressure peak in the period AB corresponds to the V wave of right atrial pressure.

97.
(a) F both receive parasympathetic fibres, the S-A node from the right vagus, the A-V node from the left vagus
(b) F Purkinje tissue is confined to the ventricles
(c) T this increases heart rate because the membrane potential reaches the threshold for firing more quickly
(d) F acetylcholine decreases the rate of diastolic depolarisation and hence the heart rate.

98.
(a) T line X represents pulmonary artery pressure
(b) T cycle length, AF, is just under one second
(c) T normal right ventricular systolic pressure is nearer 20 than 50 mmHg
(d) F it occurs at points C and H, where ventricular pressure rises to exceed pulmonary artery pressure.

99.
(a) F it occurs at point D where ventricular pressure falls below pulmonary arterial pressure
(b) T DE is longer than BC and GH
(c) T it is closed between B and E
(d) F it corresponds to the A wave which is due to atrial systole.

100. The first heart sound occurs synchronously with:
(a) closure of the aortic and pulmonary valves
(b) the P wave of the electrocardiogram
(c) a rise in atrial pressure
(d) a rise in ventricular pressure.

101. Stimulation of sympathetic nerves to the heart:
(a) increases the rate of diastolic depolarisation in cardiac pacemaker cells
(b) shifts the Starling curve of myocardial contractility (ordinate) versus end-diastolic volume (abscissa) to the right
(c) causes a rise in coronary blood flow
(d) increases the rate of conduction in the A-V node.

102. The velocity of blood flow:
(a) in the capillaries is low because the resistance they offer is high
(b) in the veins is greater than in the venules
(c) falls to about zero in the ascending aorta during diastole
(d) is greater towards the centre than towards the periphery of the stream in large blood vessels.

103. Cardiac muscle contractility:
(a) depends on an interaction between actin and myosin filaments
(b) is enhanced when the serum potassium rises above normal
(c) is lost if the bathing medium is made calcium-free
(d) is reduced by the local application of acetylcholine.

100.

(a) F it is caused by closure of the mitral and tricuspid valves

(b) F the P wave, which is caused by atrial depolarisation, precedes the first heart sound

(c) T ventricular contraction causes the A-V valves to bulge into the atria. The small rise in pressure is the C wave of the atrial pressure cycle

(d) T it is the rise in ventricular pressure which closes the mitral and tricuspid valves.

101.

(a) T this accounts for the increase in heart rate since the membrane potential reaches the threshold for firing more quickly

(b) F it shifts it to the left by increasing the force of contraction for any given end diastolic volume

(c) T a consequence of increased myocardial metabolism rather than a direct effect of sympathetic nerves on coronary vessels

(d) T as in other parts of the myocardium.

102.

(a) F it is low because the total cross-sectional area of the capillaries is large, the resistance offered by the capillaries is less than that offered by the arterioles

(b) T because the total cross-sectional area of the veins is smaller

(c) T in fact, blood flow may be momentarily reversed in the aorta in early diastole

(d) T the layer of blood in contact with the wall is slowed by the cohesive forces between it and the wall surface; velocity increases parabolically towards the centre of the vessel.

103.

(a) T as in other muscle

(b) F this diminishes cardiac contractility

(c) T extracellular calcium is essential for coupling excitation to contraction in heart muscle

(d) T however, vagal nerve stimulation has little effect on contractility since few vagal fibres impinge on ventricular muscle cells.

104. During isometric ventricular contraction:

(a) pressure is rising steadily in the aorta
(b) the entry and exit valves of the ventricle are closed
(c) the heart muscle does not expend any energy
(d) blood flow to left ventricular muscle falls.

105. In the electrocardiogram, the:

(a) QRS complex occurs synchronously with contraction of the ventricles
(b) T wave is due to repolarisation of the ventricles
(c) PR interval corresponds with the spread of depolarisation over the atria
(d) RT interval corresponds with the duration of the action potentials in ventricular myocardial cells.

106. Cardiac output:

(a) is usually expressed as the combined outputs of left and right ventricles per minute
(b) need not increase when the heart rate rises
(c) rises when the subject changes from the standing to the lying down position
(d) is reduced reflexly in a hot environment.

107. Arterioles:

(a) have a greater wall thickness to lumen diameter ratio than the arteries
(b) play a major role in regulating arterial blood pressure
(c) are the vessels in which most of the energy imparted to the blood by the ventricles is dissipated
(d) play a major role in regulating local blood flow.

104.
(a) F the pressure in the aorta is falling
(b) T hence no blood leaves the ventricle
(c) F it expends energy raising ventricular pressure to a value which opens the aortic and pulmonary valves
(d) T the blood vessels are squeezed by the contracting muscle.

105.
(a) F the complex is due to the spread of depolarisation over the ventricles; it precedes ventricular contraction
(b) T it marks the end of ventricular systole
(c) F the P wave corresponds with atrial depolarisation; the PR interval is an index of the delay experienced by the impulse in the A-V bundle
(d) T the QRS complex corresponds with depolarisation and the T wave with repolarisation.

106.
(a) F it is expressed as the output of one ventricle per minute (both the same for practical purposes)
(b) T if there is a corresponding fall in stroke volume
(c) T since the atrial filling pressure rises
(d) F it may rise to 3–5 times its resting value in a hot environment, due in part to the increase in blood flow through the skin.

107.
(a) T in arterioles the ratio is about 1:1
(b) T arterial pressure depends on cardiac output and peripheral resistance (which resides mainly in the arterioles)
(c) T the major drop in blood pressure as the blood traverses the circulation occurs in the arterioles
(d) T in a constant pressure system, flow is inversely related to resistance; since the bulk of peripheral resistance lies in arterioles, changes in their calibre cause large changes in the total resistance of the circuit.

108. In the heart, Purkinje tissue cells:

(a) are modified myocardial cells

(b) are confined to the ventricles

(c) can conduct impulses as fast as some nerves

(d) excite the myocardium of the interventricular septum before the outer walls of the ventricles.

109. In the brachial artery:

(a) the pulse wave travels towards the wrist at the same velocity as the arterial blood

(b) the pulse pressure falls with increasing age due to decreasing elasticity of the aorta

(c) pressure rises by more than 20% if the distal end of the artery is occluded

(d) blood pressure falls when the arm is raised above the head.

110. The tendency for blood flow to be turbulent is greater:

(a) in large than in small vessels

(b) when the blood is more viscous

(c) when the velocity is high

(d) as the density of the blood decreases.

111. Splitting of the second heart sound into two audible components in healthy people:

(a) can usually be heard best in the "pulmonary area", a small area in the 2nd intercostal space to the left of the manubrium sterni

(b) is usually due to pulmonary valve closure just preceding aortic valve closure

(c) is most easily heard in inspiration because this prolongs right ventricular systole

(d) tends to disappear during expiration.

108.

(a) T they are larger and paler than ventricular myocardial cells, and have a high glycogen content and sparse myofibrils

(b) T this type of specialised conducting tissue has not been identified in the atria

(c) T they conduct at about 4 metres/second which is faster than conduction in some unmyelinated nerves (C fibres)

(d) T tracts of Purkinje tissue travel down the septum to the apex before travelling up the outer walls to the base of the heart.

109.

(a) F the pulse wave travels about 10 times as fast; it is a pressure wave which does not depend on flow

(b) F systolic pressure tends to rise because the aorta is less compliant; diastolic pressure tends to fall because of the diminished elastic recoil during diastole. These changes are superimposed on an unexplained tendency for mean arterial pressure to rise with age

(c) F blood pressure in the brachial artery depends on cardiac output and *total* peripheral resistance; occlusion of one brachial artery has little effect on total resistance

(d) T by the hydrostatic equivalent of the column of blood above the heart.

110.

(a) T liability to turbulence $\propto$ diameter of vessel (D)

(b) F liability to turbulence $\propto$ 1/viscosity (η)

(c) T liability to turbulence $\propto$ velocity (V)

(d) F liability to turbulence $\propto$ density (ρ). Reynolds number (R) expresses the probability of turbulence. $R = \rho DV/\eta$.

111.

(a) T aortic valve closure can be heard over a wide area; the softer pulmonary valve closure is best heard in the pulmonary area

(b) F the aortic valve usually closes first

(c) T inspiration by decreasing intrathoracic pressure transiently improves filling of the right ventricle

(d) T for the converse of the reasons given above.

112. It may be concluded that arterioles offer more resistance to blood flow than other blood vessels because:

(a) they have thick muscular walls

(b) they have a rich sympathetic innervation

(c) they have the smallest internal diameters

(d) the pressure drop across them is greater than that across the arteries, the capillaries, the venules and veins.

113. Starling's Law of the Heart:

(a) states that the strength of myocardial contraction is a function of the initial length of the muscle fibres
(b) could explain how stroke volume may be maintained when peripheral resistance rises
(c) could explain the fall in cardiac output when a person changes from the lying to the standing position
(d) could explain why the outputs of the left and right ventricles are equal in the long term.

114. Conduction of the electrical impulse in the heart:

(a) depends on protoplasmic bridges between adjacent cardiac cells

(b) is responsible for the generation of the electrocardiogram

(c) is slowest in its transit through the atrio-ventricular node

(d) is faster in ventricular muscle fibres than in atrial muscle fibres.

112.

(a) F this suggests that the resistance may be variable; the thickness of the wall does not affect resistance

(b) F this suggests that activity in nerves could vary their resistance

(c) F the internal diameter of capillaries is smaller than that of arterioles

(d) T two-thirds of the perfusion pressure is used up getting blood through the arterioles. This indicates that their resistance is twice as great as that of all other vessels combined.

113.

(a) T Starling established this relationship using the heart-lung preparation

(b) T a rise in peripheral resistance would impede ventricular outflow and lead to stretching of the ventricular fibres

(c) T as blood pools in the dependent veins, venous return falls and the filling pressure of the heart is reduced

(d) T if one ventricle lagged behind the other, blood would accumulate and stretch its fibres, leading to an increased output.

114.

(a) F there is no protoplasmic continuity between cells. Excitation may spread from cell to cell at "tight junctions" where the membranes of adjacent cells make close contact

(b) T the moving fronts of depolarisation and repolarisation set up voltages in the body, which acts as a volume conductor, and derivatives of these voltages may be recorded from the body surface

(c) T this allows time for the atria to contract before the ventricles

(d) F ventricular conduction rate (0.4 metres/second) is slower than atrial (1.0 metres/second). The rapid excitation of ventricular muscle depends on the rapidly conducting Purkinje tissue (4.0 metres/second).

115. Rhythmic venous compression during muscular exercise in the legs:

(a) drives venous blood towards but not away from the heart
(b) is entirely responsible for venous return from the legs to the heart when the subject is standing erect
(c) increases the pressure gradient between the arteries and the veins in the feet when the subject is erect

(d) increases the formation of tissue fluid.

116. If one examines the arterial system at progressively greater distances from the heart one finds:

(a) an increasing proportion of smooth muscle

(b) a decreasing proportion of elastic tissue

(c) no change in mean blood pressure until the arterioles are reached
(d) an increasing tendency for turbulent flow.

117. The diagram below shows some related events of the cardiac cycle. In this diagram:

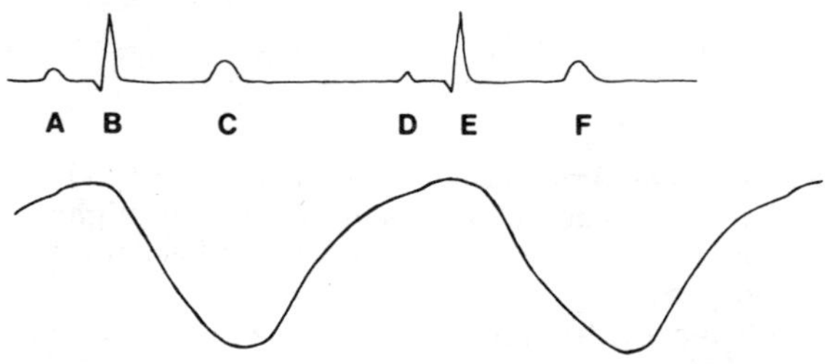

(a) B and E approximate in time to the first heart sound

(b) A and D approximate in time to ventricular repolarisation
(c) the periods BC and EF correspond to ventricular systole

(d) the lower trace could represent changes in aortic pressure.

115.

(a) T because valves prevent retrograde flow
(b) F it helps, but the pressure gradient generated by the heart ensures a flow of blood even when muscles are not active
(c) T by pumping blood out of the veins it reduces venous pressure; the increased perfusion pressure tends to increase foot blood flow
(d) F by reducing venous pressure it results in a lower capillary hydrostatic pressure.

116.

(a) T the walls of the peripheral distributing arteries consist mainly of smooth muscle
(b) T elastic tissue predominates in the walls of the aorta and proximal arteries which damp the pressure fluctuations generated by the heart
(c) F there is a small but definite pressure gradient down the arterial tree
(d) F the tendency decreases with decreasing velocity and vessel diameter.

117.

(a) T this closely follows the QRS complex of the ECG opposite
(b) F they mark atrial depolarisation
(c) T between ventricular depolarisation B, E and repolarisation C, F
(d) F it could represent changes in ventricular volume; peak arterial pressure would be seen during systole (BC and EF).

118. In the estimation of cardiac output (CO) by the indicator dilution technique:

(a) it is assumed that there is virtually complete mixing of the indicator throughout the circulation
(b) analysis of the blood samples shows a secondary rise in indicator concentration before it falls to a nearly steady level
(c) the duration, t, of the indicator curve is shorter during exercise than at rest

(d) the mean concentration, c, of the indicator curve increases as the cardiac output increases.

119. In the estimation of cardiac output using the Fick principle:

(a) it is assumed that cardiac output approximates to pulmonary blood flow
(b) the result can be obtained by multiplying pulmonary oxygen uptake (U) by the difference between arterial oxygen content (A) and the average oxygen content of venous blood entering the lungs (V)
(c) a sample from the superior (or the inferior) vena cava may be used to measure the oxygen content of venous blood entering the lungs

(d) if oxygen uptake increases fourfold, then cardiac output must have increased fourfold.

120. Intravenous infusions of:

(a) noradrenaline lead to constriction of arterioles in skeletal muscle
(b) adrenaline lead to constriction of arterioles in skeletal muscle

(c) adrenaline and noradrenaline have similar effects on arterial blood pressure

(d) adrenaline and noradrenaline have similar effects on heart rate.

118.

(a) F this is assumed in the measurement of blood volume

(b) T this is due to recirculation of the indicator, e.g. through the relatively short coronary circuit

(c) T with the higher cardiac output in exercise, the blood containing the indicator passes the sampling point more rapidly

(d) F it falls as cardiac output increases because the higher flows cause increasing dilution of the indicator dose, i; $CO = i/ct$

119.

(a) T the method measures pulmonary blood flow

(b) F pulmonary blood flow (F) = oxygen uptake divided by the arteriovenous oxygen difference; $F = U/(A\text{-}V)$

(c) F the oxygen content of blood from these vessels (and from the coronary sinus) differs widely; an adequately mixed sample must be taken from the right ventricle, or, preferably, pulmonary artery

(d) F if the arteriovenous difference (A-V) has increased simultaneously, cardiac output has increased less than fourfold.

120.

(a) T due to stimulation of alpha adrenoceptors

(b) F they lead to dilatation; adrenaline has a greater affinity than noradrenaline for beta receptors

(c) F because of its proportionately greater effect on alpha receptors and lesser effect on beta receptors, noradrenaline raises blood pressure more than adrenaline

(d) F adrenaline infusions tend to increase heart rate; noradrenaline infusions cause reflex cardiac slowing in response to the rise in arterial pressure.

121. The diagram below shows left ventricular (LV) function curves of the Frank-Starling type. In this diagram:

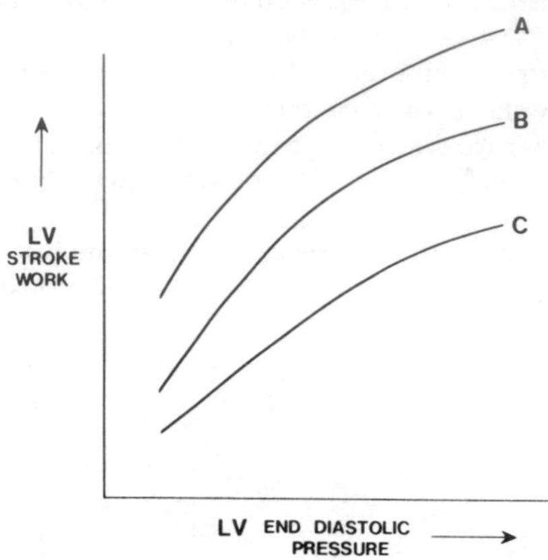

(a) appropriate units for stroke work would be millilitres per stroke

(b) end-diastolic pressure is used as an index of the degree of stretch of cardiac muscle

(c) if curve B represents function during mild sympathetic stimulation of the heart, curve C represents function during more intense stimulation

(d) if curve C represents function of the intact heart at rest, curves B and A could represent function during increasingly strenuous body exercise.

121.

(a) F the units of work are |mass| |length|, e.g. gram metres. Stroke work takes into account both stroke volume and the systemic arterial pressure which must be overcome in the ejection of the stroke volume

(b) T end-diastolic volume can also be used for this purpose

(c) F curve A corresponds to the more intense stimulation—more work is done at the same degree of stretch (same end-diastolic pressure)

(d) T there is increasing sympathetic stimulation of the heart with increasing levels of exertion.

CARDIOVASCULAR SYSTEM APPLIED PHYSIOLOGY

122. The measurements below, of digital skin temperature, were made on three persons (A, B and C) before and during a period of body heating as indicated by the vertical lines. The digits were exposed to a room temperature of 15°C throughout. The records obtained in the case of person:

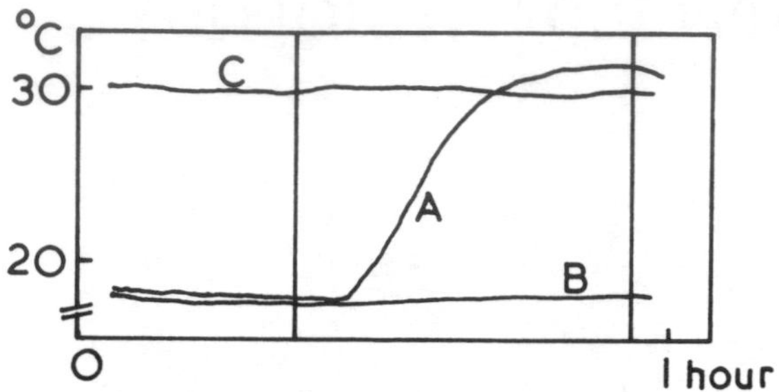

(a) A suggest release of sympathetic tone to the digits by body heating
(b) B suggest that cutting the sympathetic nerves to the digits would be unlikely to increase blood flow to the digits
(c) C are typical of someone who has recently undergone sympathectomy (cutting of the sympathetic nerves to the digits)
(d) A suggest an increased activation of alpha adrenoceptors in the digital blood vessels as body heating proceeded.

123. Sudden cessation of the heart beat (asystole):

(a) causes physical signs similar to those of ventricular fibrillation
(b) is followed by unconsciousness after an interval of 1–2 minutes
(c) should be treated by rhythmic pressure on the sternum 10–15 times per minute
(d) may be reversed by applying an electric shock of appropriate strength across the thorax.

122.

(a) T this allows vasodilatation to occur; the resulting flow of warm blood raises digital skin temperature

(b) T failure of the digits to warm during body heating suggests local vascular disease as the cause of a persistently low digital temperature

(c) T there is a permanent vasodilatation, unrelated to general body temperature; this vasodilatation tends to fade eventually

(d) F release of vasoconstrictor tone results in less release of noradrenaline at sympathetic nerve endings and hence less alpha adrenoceptor activation.

123.

(a) T in neither case is there any useful cardiac output

(b) F consciousness is lost abruptly, usually within about 5 seconds

(c) F the rate should be 60–80 per minute, the sternum being depressed about 5 cm per stroke; this may fracture ribs but the patient may recover

(d) F this is the treatment for ventricular fibrillation.

124. Fainting or syncope is typically associated with:

(a) loss of consciousness due to cerebral ischaemia
(b) a raised heart rate at the time consciousness is lost

(c) vasoconstriction in skeletal muscle

(d) cerebral damage if the individual is kept upright.

125. A sustained rise in systemic arterial pressure (hypertension) may be caused by:

(a) hypoxia due to chronic respiratory failure

(b) excessive secretion of aldosterone

(c) excessive secretion of ACTH (adrenocorticotrophic hormone)

(d) hypertrophy of the left ventricle.

126. Peripheral (e.g. hypovolaemic) differs from central (e.g. cardiac) circulatory failure, in that:

(a) underperfusion of the tissues (shock) occurs

(b) the cardiac output is usually normal
(c) the central venous pressure (CVP) is low

(d) ventricular function tends to be normal.

124.

(a) T

(b) F heart rate is usually slowed due to increased vagal activity; in some situations a rapid heart rate precedes the vagal slowing

(c) F there is vasodilatation mediated by cholinergic sympathetic vasodilator nerves; hence the name—vasovagal syndrome

(d) T this prolongs and exaggerates the cerebral ischaemia which can cause irreversible cerebral damage.

125.

(a) F in chronic respiratory failure the hypoxia causes pulmonary vasoconstriction and pulmonary hypertension

(b) T associated with excessive salt and water retention (Conn's syndrome)

(c) T this raises the level of cortisol which increases blood pressure by an obscure mechanism (Cushing's disease)

(d) F this is a consequence, not a cause of hypertension. Pressure would be normal with left ventricular hypertrophy if the peripheral resistance and cardiac output were normal.

126.

(a) F underperfusion of the tissues is common in both types of failure

(b) F cardiac output is usually reduced in both types of failure

(c) T in peripheral failure, CVP is low because of diminished venous return; in central failure CVP is raised because venous return is not cleared by the failing heart

(d) T this is a criterion of peripheral failure whereas diminished ventricular function is a criterion of central failure.

127. In atrial fibrillation:

(a) sinus arrhythmia commonly occurs

(b) the ventricular rate is usually determined by a ventricular pacemaker

(c) T waves are usually absent from the electrocardiogram

(d) the electrocardiogram shows no evidence of electrical activity in the atria.

128. In the electrocardiogram shown below:

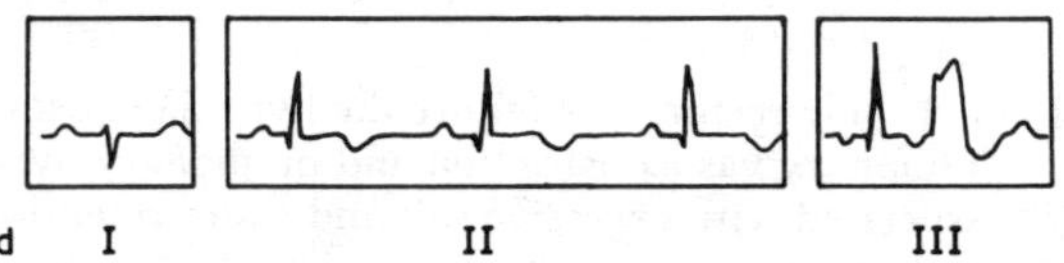

(a) sinus rhythm is present in lead II

(b) the ectopic beat in lead III is more likely to have an atrial than a ventricular origin

(c) the mean QRS vector is nearer +120° than +30° (0° is horizontal to the left; 90° is vertically downwards)

(d) the mean T vector is nearer +180° than 0°.

129. Severe hypertension may result in:

(a) ventricular hypertrophy due to an increase in the number of myocardial cells

(b) a reduction in visual acuity

(c) pulmonary oedema

(d) a marked rise in coronary arterial blood flow.

127.
(a) F in sinus arrhythmia atrial activity is normal, but impulse formation is not completely regular; in atrial fibrillation, atrial activity is markedly abnormal
(b) F the irregular ventricular beats are due to the irregular passage of electrical activity from the atria to the ventricles
(c) F the T wave of ventricular repolarisation follows each QRS complex
(d) F small rapid waves indicate the atrial fibrillation.

128.
(a) T regular QRS complexes follow regular P waves
(b) F its bizarre shape and prolonged duration indicate a ventricular ectopic origin, rather than normal spread through the conducting tissue from an atrial ectopic beat
(c) T mean QRS is slightly negative in lead I, indicating a vector to the right rather than to the left; leads II and III confirm an axis slightly to the right of the vertical
(d) F the T wave is positive in lead I, indicating an axis (vector) to the left rather than to the right.

129.
(a) F the hypertrophy in severe hypertension is due to an increase in the size of the individual fibres rather than an increase in their number (hyperplasia)
(b) T due to damage to the retina and retinal vessels
(c) T if the load on the left ventricle is so great that it causes acute left ventricular failure
(d) T because of the increased work and hence hypoxia in the myocardial tissue.

130. Auscultation of the heart provides evidence of:
(a) mitral incompetence if an early diastolic murmur is heard
(b) aortic stenosis if a systolic murmur conducted mainly into the neck vessels is heard
(c) ventricular septal defect if a loud diastolic murmur is heard
(d) mitral regurgitation if a systolic murmur is heard which is conducted into the axilla and back on the left side.

131. The electrocardiogram:
(a) shows regular but asynchronous P and QRS waves in a patient suffering from complete heart block (no conduction through A-V node)
(b) shows no P waves but regular QRS complexes in atrial fibrillation
(c) may provide evidence of right ventricular hypertrophy
(d) has an irregular saw-tooth appearance in ventricular fibrillation.

132. The jugular venous:
(a) pulse cannot usually be seen in people with normal hearts
(b) pressure is typically raised in right ventricular failure
(c) pulse is exaggerated in patients suffering from tricuspid incompetence
(d) pressure is characteristically raised in partial obstruction of the superior vena cava.

130.
(a) F mitral incompetence causes a systolic murmur; stenosis causes an early diastolic murmur
(b) T this is characteristic
(c) F the murmur is characteristically systolic
(d) T conduction indicates the direction of flow of the regurgitant blood.

131.
(a) T electrical and mechanical activity in the ventricles is independent of that in the atria in complete heart block
(b) F although the P waves are absent, the QRS complexes are irregular
(c) T e.g. if there is a prominent R wave in V_1
(d) T due to multiple abnormal pacemakers throughout the ventricle.

132.
(a) F it can be demonstrated in normal people when they lie flat or with their shoulders slightly raised
(b) T this is an important sign; venous pulsation is present
(c) T right ventricular contraction forces blood back through the atrium
(d) T an important sign of mischief in the thorax; there is distension with little or no pulsation.

133. The electrocardiograms shown below were recorded from a patient, with one year intervening between record (A) and record (B). In these records the:

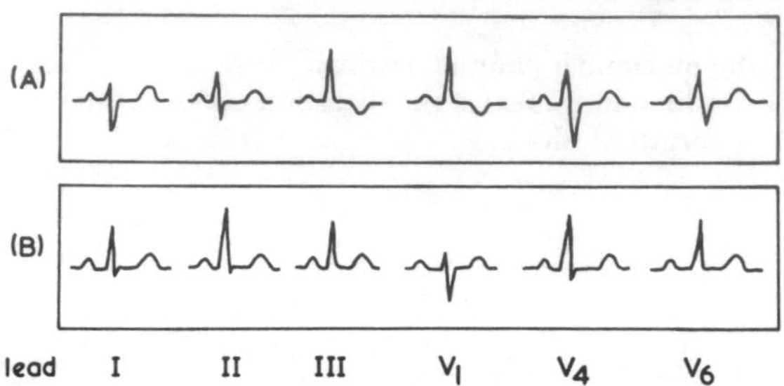

(a) QRS axis in (A) is directed to the left, rather than to the right of the vertical

(b) QRS complexes $V_{1,\ 4,\ 6}$ in (A) suggest left ventricular hypertrophy rather than right ventricular hypertrophy

(c) change from (A) to (B) suggests return towards normality

(d) QRS axis in (B) is directed downwards rather than upwards.

134. In heart failure:

(a) pitting oedema occurs earlier over the sacrum than behind the malleoli in patients who spend most of their time sitting

(b) pulmonary oedema tends to occur when pulmonary capillary pressure doubles

(c) sodium is retained to excess by the body

(d) the A-V oxygen difference is abnormally high during exercise.

133.

(a) F the net QRS in lead I is negative, indicating an axis to the right, rather than the left; this is confirmed by lead III having a large positive deflection

(b) F the R dominance in V_1 and the prominent S wave in V_4 suggest an abnormally great contribution from the right ventricle in these leads

(c) T in (B) leads V_1 and V_4 now show a normal S wave (V_1) and R wave (V_4) respectively

(d) T it is roughly 60° below the horizontal (close to the axis of lead II). The swing from right (A) to left (B) indicates return of left ventricular dominance, as do the V-lead changes.

134.

(a) F the reverse is true; oedema of cardiac origin usually occurs in the dependent parts of the body in the first place

(b) F this value is still less than plasma oncotic pressure so fluid does not accumulate in the alveoli

(c) T poor cardiac output leads to increased aldosterone secretion probably by a mechanism which is not well understood

(d) T due to increased oxygen extraction by the tissues when the heart's ability to increase cardiac output is limited.

135. Disease causing widespread destruction of lung tissue and impaired pulmonary gas exchange characteristically leads to:

(a) a rise in pressure in the pulmonary artery (pulmonary hypertension)

(b) decreased cerebral blood flow

(c) a decreased red cell count

(d) a prominent P wave in the electrocardiogram.

136. Narrowing of the coronary arteries is associated with pain which:

(a) is caused by the accumulation of pain-producing metabolites in the myocardium

(b) is typically made worse by treatment which lowers blood pressure in the systemic circulation

(c) may be precipitated by anaemia and subsequently relieved when the anaemia is treated

(d) may be elicited by sudden exposure to cold.

137. Pulmonary embolism (impaction of a clot in the pulmonary vascular bed) is a recognised complication of:

(a) prolonged bed rest

(b) a surgical operation

(c) vitamin K deficiency

(d) oral contraceptive therapy.

135.

(a) T due to increased pulmonary vascular resistance caused by the vasoconstriction induced by alveolar hypoxia and destruction of pulmonary blood vessels
(b) F retained CO_2 dilates cerebral vessels
(c) F hypoxia stimulates erythropoietin production
(d) T due to the right atrial hypertrophy which occurs once the right ventricle begins to fail.

136.

(a) T this happens when the myocardial blood flow is insufficient to clear the pain-producing metabolites at the normal rates
(b) F the pain is relieved since this reduces the work of the heart
(c) T anaemia impairs oxygen delivery while increasing myocardial demand (cardiac output is raised in anaemia)
(d) T the vasoconstriction in response to cold raises arterial pressure and thus the amount of myocardial work needed to maintain cardiac output (compare cold pressor test).

137.

(a) T immobility predisposes to venous thrombosis and hence embolism
(b) T platelet stickiness increases after the tissue damage
(c) F the tendency to thrombosis is reduced since prothrombin formation is depressed
(d) T the tendency to thrombosis is increased by an unknown mechanism.

138. The diagram below shows the relationship between cerebral blood flow and mean arterial pressure in two people, one normal, the other suffering from sustained hypertension. Normal values are indicated by the dashed lines. In this diagram:

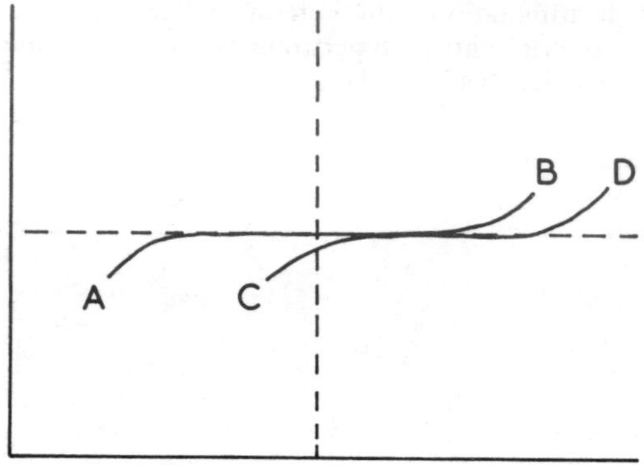

(a) the abscissa would appropriately be labelled cerebral blood flow

(b) the curve AB rather than CD would represent the normal situation

(c) there is evidence that cerebral blood flow is reduced at very high blood pressures

(d) there is evidence that lowering arterial blood pressure to 80% of normal would reduce cerebral blood flow in the person represented by CD but not in the person represented by AB.

139. Narrowing of the leg arteries is associated with:

(a) pain in the calf muscles on exercise which is relieved by rest

(b) the growth of collateral vessels which can bypass the obstruction

(c) a reactive hyperaemia which has a lower peak and a longer duration than normal, following temporary arrest of the circulation to the leg

(d) delayed healing of cuts and abrasions in the feet.

138.

(a) F the ordinate represents cerebral blood flow which is stable over a range of mean arterial pressure (cerebral autoregulation)

(b) T normally the plateau extends above and below the normal pressure, i.e. from around 60 to 160 mm Hg

(c) F at the highest pressures shown, cerebral blood flow is increased

(d) T at this pressure, flow in AB is normal, in CD it is reduced by about a quarter.

139.

(a) T exercise produces metabolites at a rate greater than that which can be cleared by the restricted blood flow

(b) T these may maintain the circulation distal to the obstruction

(c) T the arteries are incapable of transmitting the usual large and rapid increase in flow

(d) T there may also be *trophic* (nutritional) changes, e.g. loss of hair, shiny skin.

140. An arterial pulse contour:

(a) with a very sharp systolic peak signifies that ejection from the left ventricle is unusually rapid
(b) with an abnormally large pulse pressure may be caused by incompetence of the mitral valve
(c) with a slowly rising systolic phase is associated with stenosis of the aortic valve
(d) which varies greatly from beat to beat is characteristic of atrial fibrillation.

141. Murmurs (or bruits) may be detected using a stethoscope over:

(a) the larger arteries in healthy adults

(b) vessels in which there is turbulence

(c) dilatations (aneurysms) in arteries

(d) the heart in healthy young adults in early diastole.

142. An adequate supply of oxygen to the left ventricular myocardium is favoured by:

(a) the system of connections between small arterial vessels which ensure continuity of blood supply when a major coronary artery is suddenly occluded
(b) the fall in coronary vascular resistance which accompanies exercise

(c) a higher-than-average oxygen extraction from blood traversing the coronary circulation

(d) the presence in the ventricle of supporting tissue which prevents compression of the blood vessels during systole.

143. When the bundle of His is completely interrupted, the:

(a) ventricles typically contract at about 30–40 beats/min

(b) atria beat at an irregular rate
(c) PR interval on the electrocardiogram is longer than normal but remains constant from beat to beat
(d) QRS complex varies in shape from beat to beat.

140.

(a) T it may occur during exercise or anxiety or other situations where there is sympathetic stimulation of the heart

(b) F aortic incompetence may give rise to it

(c) T due to slow expulsion of blood from the left ventricle

(d) T because the volume of the ventricle at the beginning of systole, and hence the stroke output, varies widely.

141.

(a) F the circulation is so designed that it does not normally give rise to murmurs

(b) T turbulence can set up vibrations which generate sound waves

(c) T eddy currents and turbulence are set up as the blood flows into the wider section of the vessel

(d) T the physiological 3rd heart sound.

142.

(a) F though anastomoses exist, coronary arteries are functional "end arteries"; a sudden major occlusion usually causes death of an area of cardiac muscle

(b) T vasodilatation which is sufficient to meet the additional metabolic needs is caused by the fall in Po_2 in the myocardium

(c) T extraction is higher than average—venous blood from the heart is about 25% saturated with oxygen compared with the 75% or so saturation of mixed venous blood

(d) F the vessels *are* compressed.

143.

(a) T this is the usual intrinsic rate (idioventricular rhythm) of the ventricles; it may be slower

(b) F they beat at their usual regular rate (60–80 beats/min)

(c) F since the atria and ventricles are beating independently there is no fixed PR interval

(d) F there is usually a single ventricular pacemaker; hence the QRS complex is constant in shape.

144. Aortic valve incompetence typically causes:
(a) an increase in pulse pressure
(b) hypertrophy of the left ventricle
(c) an increase in myocardial blood flow
(d) eventual failure of the left ventricle.

145. A ventricular extrasystole:
(a) is associated with an abnormal QRS complex
(b) tends to be followed by a compensatory pause
(c) may fail to produce a pulse at the wrist
(d) indicates the presence of serious heart disease.

146. Pulmonary embolism (impaction in the pulmonary vascular bed of circulating clot or other matter):
(a) may cause pulmonary hypertension if the emboli are small and multiple
(b) may cause a fall in left atrial pressure
(c) usually causes death of some of the tissue whose pulmonary blood flow is blocked
(d) may cause collapse of the segment of lung supplied by the obstructed vessel.

147. Factors which increase the stiffness of arterial walls:
(a) tend to raise the systolic arterial pressure
(b) tend to raise diastolic pressure
(c) increase the peripheral resistance
(d) decrease the velocity of the arterial pulse wave.

144.
(a) T due to a very low diastolic pressure
(b) T due to the increased work needed to expel the greater stroke volume
(c) T to meet the metabolic cost of the increased work
(d) T any chamber of the heart given a persistently high work load will eventually fail.

145.
(a) T since the extrasystole arises from an ectopic focus the spread of the impulse over the ventricles is abnormal and leads to a peculiar prolonged QRS complex
(b) T since the next normal atrial impulse may reach the ventricles when they are in a refractory state—the heart then misses a beat
(c) T if it occurs early in diastole
(d) F it occurs in many normal hearts.

146.
(a) T pulmonary vascular resistance is increased
(b) T this may decrease the left ventricular filling pressure to cause a sudden massive fall in cardiac output
(c) F tissue death is unusual since enough nutrition can be supplied via the bronchial arteries
(d) T this has been attributed to loss of surfactant in the affected alveoli; a normal blood supply is needed for surfactant production.

147.
(a) T since the cardiac output is less readily accommodated in systole
(b) F diastolic pressure tends to fall due to lack of elastic recoil in diastole
(c) F stiffness of the wall does not affect resistance
(d) F a vibration travels more rapidly in a stiff than in a lax structure.

148. The *a* wave of venous pulsation in the neck:

(a) occurs just after the pulse in the carotid artery in a given cardiac cycle
(b) is exaggerated in atrial fibrillation
(c) is exaggerated in tricuspid stenosis
(d) is exaggerated in complete heart block when the P wave of the electrocardiogram falls between the QRS and T waves.

149. The diagram below shows left ventricular (LV) function curves of the Frank-Starling type. If point X on curve B represents the conditions in the normal heart at rest, then point:

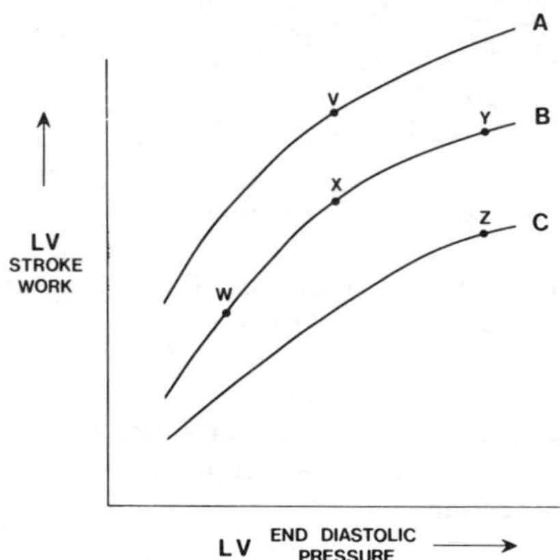

(a) Z could represent conditions in the failing ventricle
(b) Y could represent conditions in the ventricle in hypertension prior to failure
(c) Y rather than point V would represent conditions in the ventricle after administration of a drug which stimulates beta adrenoceptors
(d) V could represent conditions in the ventricle in a case of aortic valve stenosis prior to failure.

148.
(a) F since it is due to atrial systole, it precedes the arterial pulse due to ventricular systole
(b) F it is abolished in atrial fibrillation since there is no effective atrial systole
(c) T due to the more vigorous right atrial systole associated with this condition
(d) T the right atrium then contracts against a closed tricuspid valve.

149.
(a) T stroke work is subnormal; end-diastolic pressure increased
(b) F in early hypertension the left ventricle hypertrophies; stroke work is greater than normal at a given filling pressure (curve A)
(c) F such a drug, e.g. isoproteronol (isoprenaline), mimics sympathetic stimulation and moves the ventricular curve upwards and to the left
(d) T stroke work is considerably increased, end-diastolic pressure little increased.

150. Left ventricular failure tends to cause:

(a) a rise in lung compliance

(b) a presystolic murmur heard over the heart

(c) attacks of breathlessness when the patient lies down

(d) a decrease in ventricular end-diastolic pressure.

151. The diagram below shows some events of the cardiac cycle in an abnormal heart with a diastolic murmur (narrow vertical lines). The diagram shows:

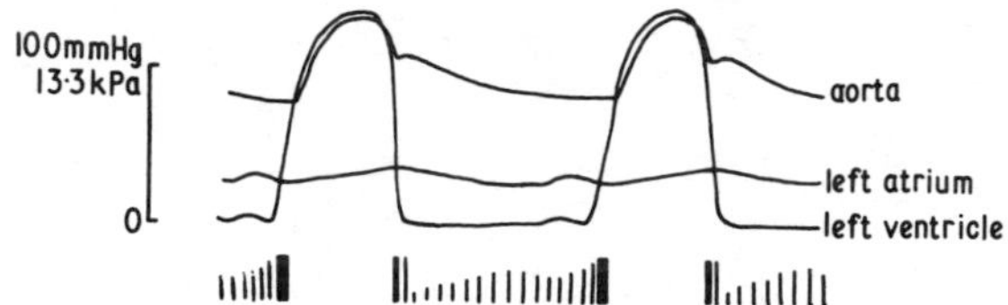

(a) an abnormally high pressure gradient across the mitral valve when it is open
(b) typical features of mitral valve incompetence
(c) features which suggest sinus rhythm rather than atrial fibrillation

(d) evidence of aortic regurgitation.

152. Following obstruction of a major coronary artery:
(a) reflex vagal inhibition of the heart may so slow the heart rate that myocardial damage is extended
(b) atrial fibrillation is a commoner consequence than ventricular fibrillation

(c) ST depression is commonly seen in a Lead II electrocardiogram
(d) a rise in body temperature is common.

150.
(a) F it is decreased as the congestion of the pulmonary vessels with blood increases the stiffness of the lungs
(b) T it is due to turbulence caused by the more forceful left atrial contraction attempting to force blood into the ventricle. The altered heart sounds are referred to as a *presystolic gallop rhythm*
(c) T the sudden increase in venous return may so overload the left ventricle that severe pulmonary congestion and perhaps oedema occur (cardiac asthma)
(d) F it tends to be increased, a compensatory mechanism which permits the ventricle to accomplish more stroke work.

151.
(a) T pressure in the left atrium is normally almost identical to that in the left ventricle during diastole
(b) F the features are typical of mitral stenosis
(c) T the presystolic peak in atrial pressure and accompanying accentuation of the murmur show co-ordinated atrial contraction
(d) F this would cause a greater arterial pulse pressure.

152.
(a) T the reflex slowing may result from afferent impulses from the damaged myocardium
(b) F ventricular fibrillation is the commoner complication, due to multiple abnormal pacemakers being set up in the tissue surrounding the injured muscle
(c) F ST elevation is the usual change
(d) T due to release of pyrogens from the damaged tissue.

153. The diagram below shows left ventricular and aortic pressures, with diagrammatic heart sounds and murmur, in an abnormal heart. It can be deduced that:

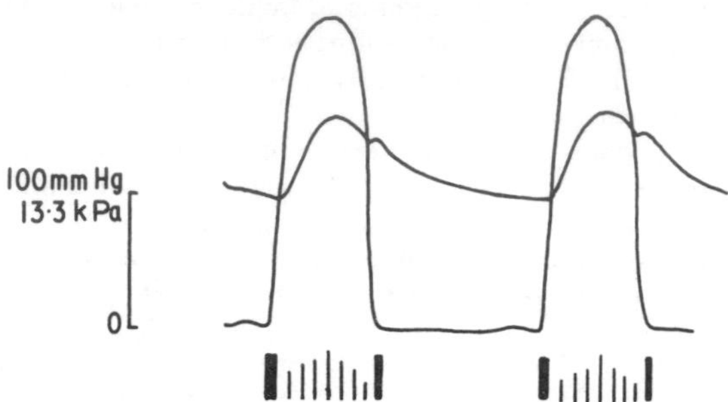

(a) aortic systolic pressure is nearer 250 than 150 mm Hg
(b) the pressure gradient across the aortic valve in systole reaches a maximum nearer 50 than 10 mm Hg
(c) the murmur would coincide in time with the RT period of the electrocardiogram
(d) the features are those of aortic valve stenosis.

154. In a previously healthy person, sudden obstruction of:
(a) an internal carotid artery results in death of brain tissue

(b) a renal artery results in death of renal tissue
(c) the femoral artery results in gangrene of the toes

(d) the hepatic portal vein results in death of hepatic tissue.

155. In the electrocardiogram below:

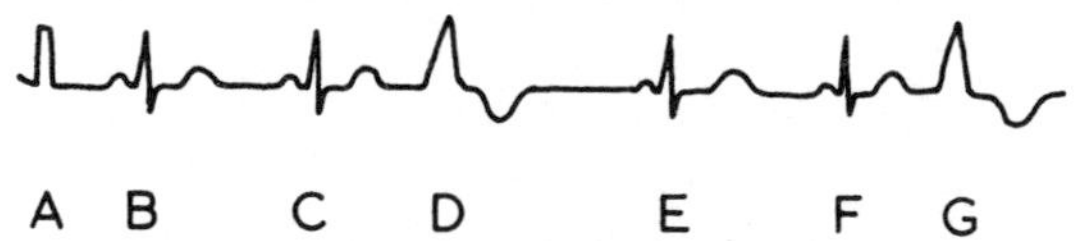

(a) B, C, E, F, mark typical sinus beats
(b) the complex at A is a typical atrial ectopic beat
(c) the complexes D, G are typical ventricular ectopic beats

(d) DE marks a compensatory pause.

153.

(a) F it is about 160 mm Hg
(b) T it is about 75 mm Hg (normally it is negligible)

(c) T it is a systolic murmur

(d) T the large pressure drop across the aortic valve suggests a great increase in aortic valve resistance.

154.
(a) F sufficient collateral supply is obtained via the circle of Willis
(b) T necrosis occurs firstly in the cortical region
(c) F the opening of collateral vessels can maintain the viability of all the leg tissues if the blood vessels of the leg are otherwise healthy
(d) F the hepatic artery can maintain hepatic viability

155.

(a) T sinus rhythm is the basic rhythm present
(b) F it is a voltage calibration, unlike any ectopic beat
(c) T they are clearly different from and longer than the normal QRS complex; their similarity to each other suggests a common focus of origin
(d) T this commonly, but not invariably, follows an ectopic beat.

RESPIRATORY SYSTEM BASIC PHYSIOLOGY

156. In a normal man breathing quietly at rest the partial pressure of:

(a) carbon dioxide in alveolar air is about twice that in room air

(b) carbon dioxide in mixed venous (pulmonary arterial) blood is greater than in alveolar air

(c) water vapour in alveolar air is less than half that of alveolar carbon dioxide

(d) oxygen in expired air is greater than in alveolar air.

157. As blood passes through systemic capillaries:

(a) its pH rises

(b) bicarbonate ions pass from the red cells to the plasma

(c) the concentration of chloride ions in the red cells falls

(d) its oxygen dissociation curve shifts to the right.

158. The respiratory centre:

(a) is situated in the hypothalamus

(b) sends out regular bursts of impulses to the inspiratory muscles during quiet respiration

(c) sends out regular bursts of impulses to expiratory muscles during quiet respiration

(d) is inhibited during swallowing and vomiting.

156.

(a) F alveolar P_{CO_2} is around 40 mm Hg (5.3 kPa); room air P_{CO_2} is about 0.2 mm Hg (0.03 kPa)

(b) T this permits passive diffusion of carbon dioxide from pulmonary capillaries to alveolar air

(c) F the average values are 47 mm Hg (6.3 kPa) for water vapour; 40 mm Hg (5.3 kPa) for CO_2

(d) T since expired air consists of alveolar air diluted with dead-space air.

157.

(a) F it falls largely because CO_2 diffusing into the blood forms carbonic acid which dissociates to give free hydrogen ions

(b) T the reaction $CO_2 + H_2O \rightleftharpoons H_2CO_3 \rightleftharpoons H^+ + HCO_3^-$ occurs mainly in the red cells because they contain carbonic anhydrase; most of the bicarbonate so formed passes into the plasma

(c) F chloride ions move from plasma to red cells to maintain electrical neutrality as bicarbonate ions move in the opposite direction ("chloride shift")

(d) T due to the rise in P_{CO_2} and H^+ concentration.

158.

(a) F the cell bodies of the neurones responsible for respiratory movements are found in the reticular substance of the medulla oblongata

(b) T this activity is responsible for increasing the dimensions of the thoracic cage

(c) F expiration is passive under resting conditions

(d) T this reflex inhibition is essential to prevent aspiration of food or vomitus.

159. The carotid bodies:

(a) have, per unit mass, a blood flow rate similar to that of brain tissue
(b) are more influenced by arterial Po_2 than by arterial oxygen content
(c) are stimulated by a rise in blood hydrogen ion concentration
(d) and the aortic bodies are entirely responsible for stimulation of ventilation in response to hypoxia.

160. The surfactant material lining the lung alveoli:

(a) decreases the surface tension of the alveolar fluid
(b) increases the compliance of the lungs
(c) has progressively less effect, the more the lungs are inflated
(d) is decreased when pulmonary blood flow is interrupted.

161. Comparing a healthy person of 70 with a healthy person of 20 (of the same sex and height), the older person would tend to show:

(a) a smaller ratio of residual volume: vital capacity
(b) lung airways which collapse and close at an earlier stage of expiration
(c) a smaller ratio of forced expiratory volume in one second: vital capacity (FEV_1:VC)
(d) a reduction in arterial oxygen saturation nearer 10% than 5%.

159.
(a) F per unit mass, they have the greatest flow yet discovered in the body
(b) T hence they are not influenced by a low haemoglobin level or CO poisoning
(c) T increased acidity of blood stimulates ventilation
(d) T when these bodies are denervated, hypoxia causes depression of ventilation.

160.
(a) T it reduces the tendency of the surface forces to collapse the lungs
(b) T since it permits them to be more easily inflated
(c) T this tends to prevent over-inflation of alveoli
(d) T this may account for the local lung collapse (atelectasis) following a pulmonary embolus.

161.
(a) F residual volume increases with age whereas vital capacity decreases
(b) T this is an important cause of the increased residual volume
(c) T the ratio falls with age, probably in part due to loss of elastic tissue which normally holds open small airways during forced expiration (see also (b) above)
(d) F although Po_2 falls to this extent, the plateau of the oxygen dissociation curve ensures a much smaller fall in saturation.

162. The diagram below indicates some events during two respiratory cycles, where I = inspiration and E = expiration. In the second cycle, tidal volume was three times that in the first cycle. Expiration was not forced. It can be concluded that:

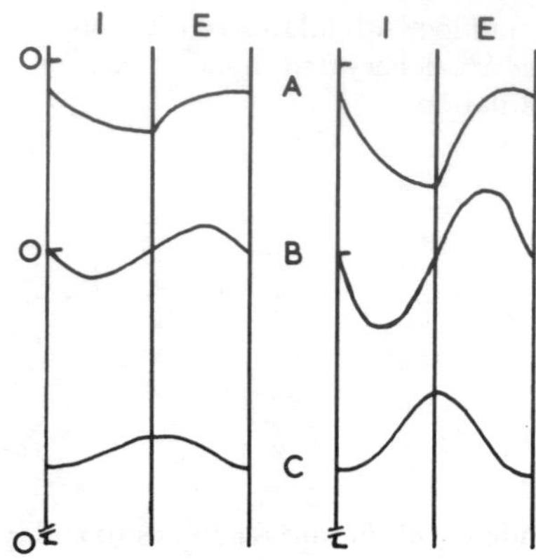

(a) record B, rather than records A or C, corresponds to intrapleural pressure

(b) record C, rather than records A or B, corresponds to intrapulmonary pressure

(c) record B, rather than records A or C, corresponds to the rate of gas flow into or out of the lungs, taking inward flow as negative

(d) compliance would be increased if record A showed smaller changes (other records unchanged).

163. Carbon dioxide:

(a) is carried on the haemoglobin molecule as carboxyhaemoglobin

(b) in the blood increases the oxygen-binding power of haemoglobin

(c) when breathed at a concentration of 20% in oxygen is a powerful respiratory stimulant

(d) in free plus combined form is present in arterial blood in greater amount than is oxygen in free plus combined form.

162.

(a) F intrapleural pressure is negative throughout the cycle (in the absence of forced expiration) and is around a minimum at the end of inspiration (record A)

(b) F intrapulmonary pressure reaches a minimum around mid inspiration and a maximum around mid expiration (record B)

(c) T the flow record closely follows the intrapulmonary pressure record, since flow is directly related to pressure gradient between lungs and atmosphere

(d) T the ratio of volume change for a given intrapleural pressure change would be increased.

163.

(a) F carboxyhaemoglobin (CO Hb) is the result of the interaction of haemoglobin and carbon monoxide. CO_2 is carried as carbaminohaemoglobin (CO_2 Hb)

(b) F it decreases it; the O_2 dissociation curve is shifted to the right

(c) F 5% CO_2 is a respiratory stimulant; over 10% it causes respiratory depression, and 20% would lead to coma

(d) T about 50 ml CO_2 versus about 20 ml O_2.

164. In normal lungs:
(a) the volume of air ventilating the alveoli per minute is greater than the volume of blood perfusing the lung capillaries per minute
(b) in the erect posture the ventilation/perfusion ratio increases from base to apex
(c) oxygen transfer can always be explained by passive diffusion
(d) dead space volume may increase by more than half during a maximal inspiration.

165. The rate of ventilation of the alveoli will be altered when a person changes from breathing room air to breathing:
(a) 21% oxygen and 79% nitrogen
(b) 17% oxygen and 83% nitrogen
(c) 2% carbon dioxide and 98% oxygen
(d) a gas mixture which causes alveolar P_{CO_2} to rise by 10%.

166. Contraction of the smooth muscle in the respiratory tract occurs in response to:
(a) irritation of the bronchial mucosa
(b) stimulation of local beta adrenoceptors
(c) a decrease in the P_{CO_2} in the bronchial air
(d) a cold stimulus to the bronchial mucosa.

167. During the initial part of inspiration:
(a) intrapulmonary pressure falls
(b) intrathoracic pressure rises
(c) intra-abdominal pressure rises
(d) the partial pressure of oxygen in the dead space rises.

164.

(a) F the ventilation/perfusion ratio is about 0.8 (alveolar ventilation, 4.0 litres; pulmonary capillary blood flow, 5.0 litres)

(b) T due to hydrostatic factors perfusion of pulmonary capillaries diminishes towards the apex of the lungs

(c) T pulmonary venous oxygen tension never exceeds mean alveolar oxygen tension

(d) T the trachea and bronchi expand as the lungs expand.

165.

(a) F this is the approximate composition of air

(b) F oxygen lack by itself does not affect breathing until inspired oxygen falls to around 15%

(c) T the stimulating effect of CO_2 is little antagonised by the excess oxygen

(d) T this is enough to double the volume breathed per minute.

166.

(a) T via a parasympathetic reflex

(b) F this leads to relaxation

(c) T this tends to limit ventilation of overventilated lung segments

(d) T normally the upper respiratory tract warms the inspired air and prevents this reflex from occurring.

167.

(a) T this is essential if air is to enter

(b) F it must fall if the lungs are to expand

(c) T due to descent of the diaphragm

(d) T inspired air replaces air which came from the alveoli in the previous expiration.

168. The diagram below shows some relationships between lung volume (increasing upwards) and oesophageal pressure during normal tidal volume. In this diagram:

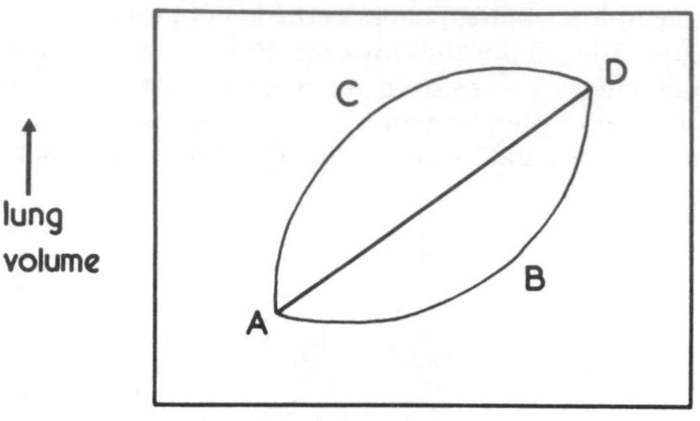

(a) the intraoesophageal pressure would be equal to atmospheric pressure at point A

(b) the changes during the respiratory cycle would follow the path ABDC
(c) the slope of the line AD is a measure of compliance
(d) the width of the loop CB is inversely related to airway resistance.

169. Compliance of the lungs is greater:
(a) in the tidal volume range, than in the inspiratory reserve volume range of ventilation
(b) at age 30 than at age 10

(c) than compliance of the lungs and thorax
(d) when filled with normal saline than when filled with air.

170. An individual who ascends from sea level to 6 000 metres (where the atmospheric pressure is halved) is likely to develop:
(a) an increase in pulmonary ventilation
(b) a fall in arterial Po_2
(c) a rise in arterial pH
(d) a rise in cerebral blood flow.

168.

(a) F intraoesophageal pressure is similar to intrapleural pressure which is negative with respect to atmospheric pressure at the beginning of a normal inspiration

(b) T since the volume changes tend to lag behind the pressure changes the relationship is a hysteresis loop

(c) T compliance is volume change for a given pressure change

(d) F the greater the hysteresis the greater the airway resistance.

169.

(a) T it is maximal in this range

(b) T there is normally a several-fold increase; the volume of gas shifted per unit pressure range is much smaller in a smaller person

(c) T it is nearly twice as great

(d) T since there are no surface forces being exerted by the fluid normally lining the alveolar membranes.

170.

(a) T due to stimulation of chemoreceptors by oxygen lack

(b) T the low blood P_{O_2} is the effective respiratory stimulant

(c) T due to the reduced P_{CO_2} caused by hyperventilation

(d) F the low blood P_{CO_2} causes cerebral vasoconstriction.

171. Inspiration:

(a) increases the venous return to the heart

(b) is assisted by surface tension forces in the alveoli

(c) requires less muscular effort than expiration during quiet breathing

(d) begins when intrapleural pressure falls below atmospheric pressure.

172. Contraction of the diaphragm:

(a) is essential to provide enough pulmonary ventilation to support life

(b) leads to protrusion of the anterior abdominal wall

(c) causes an increase in the pressure gradient between the inside of the lungs and the intrapleural space

(d) ceases if the spinal cord is severed at the seventh cervical segment.

173. The diffusing capacity or gas transfer factor of the lungs:

(a) may be expressed as volume per unit time per unit pressure gradient

(b) is greater for oxygen than for carbon dioxide

(c) increases in exercise

(d) is unaffected if one lung is excised.

174. Tidal air in a resting subject:

(a) has an average volume of around three times the dead space

(b) has a reciprocal relationship with the respiratory rate

(c) has a similar volume to alveolar air volume

(d) leaves the body saturated with water vapour.

171.

(a) T by decreasing and increasing intrathoracic and intra-abdominal pressure respectively

(b) F it must overcome these forces

(c) F expiration is passive in quiet breathing

(d) F intrapleural pressure is usually lower than atmospheric pressure even in expiration; inspiration occurs when *intrapulmonary* pressure falls below atmospheric pressure.

172.

(a) F sufficient ventilation to support life can be achieved by the activity of intercostal muscles

(b) T by increasing intra-abdominal pressure

(c) T the intrapleural pressure falls, causing the lungs to expand and air to flow into the lungs until intrapulmonary pressure returns to atmospheric

(d) F the phrenic nerve which arises from cervical segments 3–5 is responsible for diaphragmatic movements.

173.

(a) T the rate of passive diffusion of a gas through a diffusion barrier is directly proportional to the pressure gradient

(b) F it is much greater for CO_2 since CO_2 is a much more diffusible gas, mainly due to its much greater solubility

(c) T it may treble; opening up and stretching of alveoli and capillaries increases the area and decreases the distance for diffusion

(d) F half the area available for transfer has been removed.

174.

(a) T it has a volume around 500 ml; average dead space volume is about 150 ml

(b) T to maintain a constant minute volume

(c) F it has a small volume (0.5 litre) relative to alveolar air volume (3.0 litres); tidal exchange merely "washes" the large alveolar reservoir

(d) T there is a daily water loss from the respiratory tract of about 500 ml.

175. The residual volume:

(a) is the gas remaining in the lungs at the end of a full expiration
(b) averages 3 – 5 litres in an adult
(c) may be measured by an indicator dilution technique
(d) decreases with age.

176. A rise in arterial carbon dioxide tension:

(a) occurs during moderate exercise
(b) stimulates respiration via peripheral chemoreceptors
(c) stimulates respiration via central chemoreceptors
(d) causes a reflex fall in blood pressure.

177. An increase in ventilation occurs when the:

(a) plasma bicarbonate level falls

(b) subject goes to sleep

(c) pH of the CSF falls
(d) blood adrenaline level rises.

178. When a person hyperventilates voluntarily so that his minute volume is trebled, the:

(a) magnitude of the negative charge on the plasma proteins decreases
(b) level of ionised calcium in the blood rises

(c) alveolar Po_2 trebles

(d) oxygen saturation of arterial blood rises by about 10%.

175.
(a) T this is its definition
(b) F it averages 1–1.5 litres
(c) T helium may be used as an indicator
(d) F it increases on average by half to one litre between the ages of 20 and 70, partly due to loss of elastic recoil in the lungs.

176.
(a) F in many instances arterial P_{CO_2} falls during exercise
(b) T the carotid and aortic bodies
(c) T the central effect predominates
(d) F it causes a reflex rise in pressure.

177.
(a) T the fall in bicarbonate causes a metabolic (non-respiratory) acidosis which stimulates ventilation
(b) F during deep sleep, ventilation falls and the alveolar and hence arterial P_{CO_2} rises
(c) T this is how the central effect of a rise in P_{CO_2} is mediated
(d) T the mechanism is obscure but may be a direct effect on the respiratory centre (injections of adrenaline into the carotid artery increase ventilation).

178.
(a) F it increases because blood pH moves further away from the isoelectric point of plasma proteins (around pH 5)
(b) F it falls because the increased negative charge on the protein binds more calcium
(c) F alveolar P_{O_2} is about 100 mm Hg (13.3 kPa) and room air P_{O_2} is about 150 mm Hg (20 kPa). Hyperventilation could not increase alveolar P_{O_2} above this value
(d) F with normal ventilation the blood is almost fully saturated (95–98%).

179. If peripheral chemoreceptor function is lost:
(a) a 75% fall in arterial Po_2 will not appreciably alter ventilation

(b) a 10% rise in Pco_2 will not appreciably alter ventilation

(c) ventilation will not increase in exercise

(d) the subject is less able to adapt to life at high altitude.

180. In the pulmonary vascular bed:
(a) vascular resistance is about ¾ of that in the systemic circuit

(b) the flow per minute is similar to that in the systemic vascular bed
(c) about half of the blood volume is accommodated

(d) flow may increase several-fold with little change in mean pulmonary artery pressure.

181. Carbon dioxide is carried in the blood:
(a) in combination with haemoglobin
(b) in physical solution in plasma
(c) in combination with plasma proteins
(d) mainly as bicarbonate.

182. A shift of the oxygen dissociation curve of blood to the right:
(a) occurs in the pulmonary capillaries
(b) is favoured by a rise in temperature
(c) favours the passage of oxygen from blood to tissues

(d) occurs when fetal blood is replaced by adult blood.

179.

(a) F respiration will be depressed due to the central effect of oxygen lack

(b) F changes in P_{CO_2} stimulate respiration mainly by a central effect; this rise will cause marked stimulation

(c) F peripheral chemoreceptors are not essential for the ventilatory response to exercise

(d) T peripheral chemoreceptors are essential for the reflex increase in ventilation with hypoxia.

180.

(a) F it is about $^1/_6$ since the pressure drop across the pulmonary circuit is about $^1/_6$ of that across the systemic circuit

(b) T if it were not, blood would accumulate in one or other bed

(c) F the pulmonary bed contains about a quarter of the blood volume

(d) T in exercise, large increases in flow can occur with minimal pressure changes, presumably because of passive dilatation of the resistance vessels.

181.

(a) T in the carbamino form; $R{-}NH_2{+}CO_2{=}R{-}NH{\cdot}COOH$

(b) T this is responsible for the P_{CO_2}

(c) T again in carbamino form

(d) T 80–90% is in this form.

182.

(a) F the shift is in the other direction

(b) T as occurs in active tissues

(c) T since more oxygen is displaced from haemoglobin for a given fall in P_{O_2}

(d) T adult blood has less affinity for oxygen than has fetal blood at tissue oxygen tensions.

183. The work of breathing:
(a) is inversely related to lung compliance
(b) increases during exercise
(c) is independent of the rate of breathing for a given alveolar ventilation rate
(d) is decreased when the subject lies down.

184. The compliance of the lungs and chest:
(a) may be expressed as a volume change per unit change in pressure
(b) is greatest over the range of chest movement found in quiet breathing
(c) is increased by the surface tension of the fluid lining the alveoli
(d) will be changed if the line relating volume to pressure is displaced to a parallel position.

185. Respiratory dead space:
(a) saturates the air with water vapour before it reaches the alveoli
(b) removes all particulate matter from the air before it reaches the alveoli
(c) decreases when the blood adrenaline level rises
(d) decreases during coughing.

186. Vital capacity:
(a) is the volume of air expired from full inspiration to full expiration
(b) increases gradually with age in adults
(c) is greater in men than in women of the same age and height
(d) is equal in volume to the sum of the inspiratory and expiratory reserve volumes.

183.
(a) T the more compliant the lungs, the less work is required to inflate them
(b) T since the rate and depth of ventilation are increased
(c) F it is minimal around the normal rate of breathing
(d) F on lying down, chest compliance falls for a number of reasons including an increased pulmonary blood volume and increased pressure of abdominal contents on the diaphragm.

184.
(a) T the normal value is about 0.1 litre/cm H_2O (1 litre/kPa)
(b) T hence work required to ventilate lungs is at a minimum in this range
(c) F the surface tension of alveolar fluid decreases compliance; the effect is normally reduced by the presence of surfactant in the alveoli
(d) F the slope of the line indicates compliance and will be unchanged.

185.
(a) T this prevents drying of the lungs
(b) F particles less than 2.0 μm can enter the alveoli and are removed by lymphatics. Dead space filters out larger particles
(c) F it increases—adrenaline dilates the lower airways by relaxing smooth muscle and the upper airways by vasoconstriction which "shrinks" the nasal mucosa
(d) T the constriction of the respiratory tract allows the air to be expelled at high velocity.

186.
(a) T this is a definition of vital capacity
(b) F it falls on average by about a litre between the ages of 20 and 70
(c) T by half to one litre
(d) F it is equal to these plus the tidal volume.

187. In the pulmonary circulation:

(a) carbonic anhydrase catalyses the breakdown of carbonic acid to water and CO_2

(b) hydrogen ions dissociate from haemoglobin

(c) blood flow tends to be diverted from poorly ventilated to well ventilated alveoli

(d) a 25% drop in the oxygen saturation of blood leaving the lungs will occur if the alveolar Po_2 falls by 25% from the normal level.

188. An "oxygen debt" is:

(a) the amount of oxygen in excess of the resting metabolic needs that must be consumed after completion of exercise

(b) built up because the pulmonary capillary walls limit the uptake of oxygen at high rates of oxygen consumption

(c) related to the fact that skeletal muscle can function temporarily in the absence of oxygen

(d) associated with a rise in blood lactate.

189. In the diagram below of blood oxygen dissociation curves:

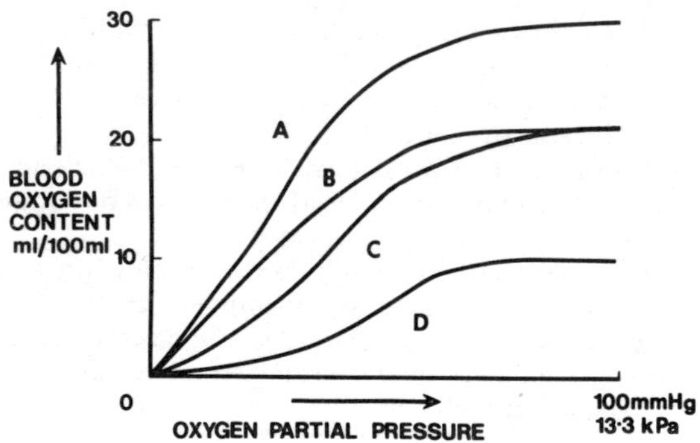

(a) curve A is consistent with blood having a haematocrit (packed cell volume) nearer 60% than 40%

(b) curve C could represent blood with a similar haemoglobin concentration to B at a higher Pco_2

(c) blood represented by curve A has an oxygen saturation more than twice that of curve D at the plateau phase

(d) blood represented by curve D is typical of someone who has lived for some months at an altitude where barometric pressure is half that at sea level.

187.

(a) T this enzyme speeds the reaction $H_2CO_3 \rightleftharpoons CO_2 + H_2O$ in either direction

(b) T in its buffering action in the systemic capillaries haemoglobin combined with these hydrogen ions

(c) T due to the constrictor action of a low alveolar Po_2 on pulmonary vessels

(d) F the drop in oxygen saturation is much less due to the plateau in this range of the oxygen dissociation curve.

188.

(a) T this is its definition

(b) F oxygen uptake by the lungs is not a limiting factor

(c) T the oxygen debt built up during anaerobic metabolism must be paid off after exercise

(d) T during anaerobic metabolism muscle glycogen is broken down into lactic acid.

189.

(a) T oxygen content at Po_2 100 mm Hg is about 30 ml/100 ml. This is 50% above average, corresponding to a haematocrit above 60%

(b) T an increase in Pco_2 displaces the curve to the right

(c) F the plateau indicates that the blood in *both cases* is almost fully saturated

(d) F curve D is consistent with a reduced haemoglobin concentration since its oxygen content is only 10 ml/100 ml at Po_2 100 mm Hg (13.3 kPa).

190. In the diagram below of blood carbon dioxide dissociation curves:

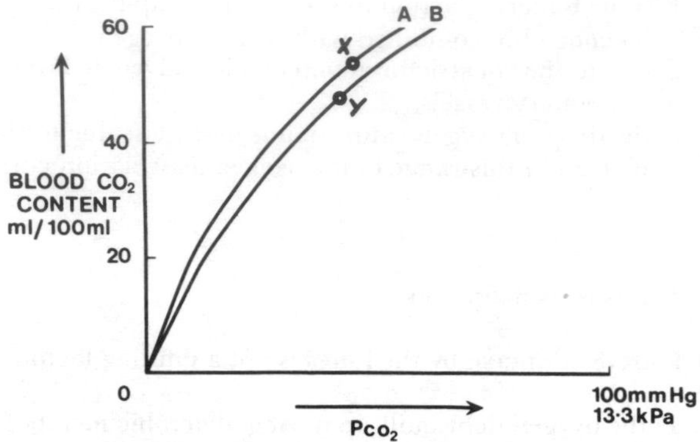

(a) a fall in P_{O_2} would tend to replace curve B by curve A

(b) if point X represents the situation at the venous end of systemic capillaries, then point Y could represent the situation when the same blood is at the venous end of pulmonary capillaries

(c) a rise in P_{CO_2} would tend to replace curve B by curve A

(d) the decrease in the slope of the curves as P_{CO_2} rises is related to the saturation of plasma with CO_2 as P_{CO_2} rises.

190.

(a) T deoxygenated blood can carry more CO_2 than oxygenated blood

(b) T in the lungs the blood oxygen saturation rises (curve A to curve B) and the CO_2 content falls (point X to point Y)

(c) F it would merely shift the position on a given curve

(d) F plasma does not become saturated with CO_2; the CO_2 content remains proportional to the P_{CO_2}.

RESPIRATORY SYSTEM APPLIED PHYSIOLOGY

191. The diagram below shows two blood oxygen dissociation curves. A represents the oxygen partial pressure in normal alveoli, H the lowered alveolar oxygen pressure in hypoxic lungs (due to high altitude or pulmonary disease) and V the mixed systemic venous oxygen pressure in the person suffering from hypoxia. In this diagram:

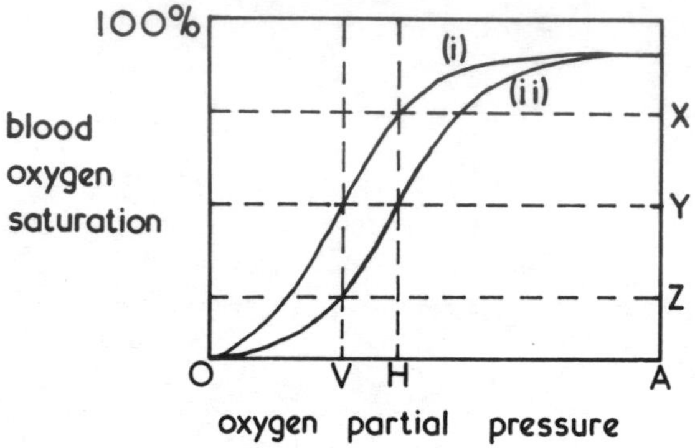

(a) if (i) is a normal person's curve, then (ii) is the hypoxic person's curve, rather than *vice versa*
(b) other things being equal, the blood represented by curve (i) would have the higher red cell level of 2,3-diphosphoglycerate (2,3-DPG)
(c) the oxygen saturation of blood leaving the hypoxic lungs would be lower in the case of curve (ii)
(d) the oxygen extracted by the tissues is similar for both curves, other things being equal.

192. The oxygen content of mixed venous blood (in the pulmonary artery) is:
(a) lower at rest than during generalised muscular exercise

(b) higher in a cool environment than in a warm environment

(c) lower than normal in anaemia

(d) lower than normal during exercise in circulatory failure.

191.

(a) T the hypoxic person's curve is displaced to the right

(b) F 2,3-DPG shifts the curve to the right

(c) T this is apparent from the graph, and is a disadvantage produced by the increased 2,3-DPG

(d) T XY = YZ, i.e. decreased alveolar uptake (X→Y) is compensated by increased delivery to the tissues (Y→Z).

192.

(a) F during exercise the blood draining the muscles has a lower oxygen content

(b) F in the cool environment, skin blood flow is lower and hence oxygen extraction relatively greater; if there is increased muscle tone or shivering, blood from muscles will also tend to lower the mixed venous oxygen content

(c) T due to the lower oxygen carrying power of the blood and the greater oxygen extraction in the blood perfusing the tissues

(d) T it is lower than normal due to the increase in oxygen extraction that occurs when the cardiac output cannot rise to meet the metabolic needs of the exercise.

193. Bronchial asthma due to allergic factors is likely to be relieved by:

(a) a drug which mimics the action of acetylcholine
(b) a beta adrenoceptor antagonist
(c) drugs such as sodium chromoglycate which stabilise mast cell membranes

(d) a drug with glucocorticoid action.

194. Air in the pleural cavity (pneumothorax) causes:

(a) a reduction in the residual volume

(b) abnormal dullness to percussion of the affected side
(c) less obvious rib contours on the affected side

(d) a reduced vital capacity.

195. A patient suffering from chronic respiratory failure (raised PCO_2, reduced PO_2):

(a) shows an increased respiratory sensitivity to carbon dioxide
(b) does not increase his ventilation in response to oxygen lack
(c) should be given 100% oxygen on admission to hospital

(d) must have been given oxygen therapy recently if his P_{CO_2} is found to be 150 mm Hg (20 kPa).

196. In emphysema:

(a) narrowing of airways is typically due to loss of pulmonary elastic tissue rather than smooth muscle spasm
(b) hypoxia may occur due to a fall in the overall ventilation/perfusion ratio
(c) the vital capacity is increased

(d) the ratio of the forced expiratory volume in one second ($FEV_{1.0}$) to the vital capacity is abnormally low.

193.

(a) F acetylcholine causes bronchoconstriction
(b) F stimulation of beta receptors dilates bronchi
(c) T drugs such as sodium chromoglycate are thought to reduce histamine release from mast cells in allergic situations
(d) T by suppressing allergic effects. Though the mechanism by which they relieve asthma is unclear, these drugs decrease inflammation, potentiate the effects of β-receptor agonists and have a membrane stabilising effect on mast cells.

194.
(a) T without negative intrapleural pressure the lung is free to follow its tendency to collapse
(b) F the air present causes increased resonance
(c) T the chest wall is not pulled in on the affected side by the elastic recoil of the lung
(d) T since the lungs cannot be fully expanded.

195.

(a) F these patients have impaired sensitivity to CO_2
(b) F oxygen lack may be the main stimulus to breathe
(c) F this may precipitate coma due to respiratory depression by removing the stimulus of hypoxia; 25–30% O_2 would be more appropriate
(d) T breathing air, $P_{CO_2} + P_{O_2} = 140$ mm Hg (18.7 kPa) as a rough rule. (P_{H_2O} and P_{N_2} are relatively fixed in alveolar air.)

196.
(a) T elastic fibres normally hold small airways open (guy-rope effect) especially during forced expiration
(b) T emphysema may cause local upsets of the ventilation/perfusion ratio in either direction
(c) F it is reduced due to small airway closure trapping a high residual volume
(d) T the $FEV_{1.0}$ is reduced even more than the vital capacity.

197. Complete obstruction of a bronchus results in:

(a) collapse of the alveoli supplied by that bronchus

(b) a rise in the intrapleural pressure on the affected side

(c) an increase in the physiological dead space

(d) vasodilatation in the alveoli supplied by the bronchus.

198. A shift of the oxygen dissociation curve of blood (with partial pressure of oxygen as the abscissa) to the left of the normal position:

(a) tends to reduce the oxygen content of circulating blood at a given Po_2

(b) tends to impair delivery of oxygen at normal tissue Po_2 levels

(c) is typically found in anaemia

(d) is typically found in blood which has been stored for several weeks.

199. Results are given below of a person's vital capacity (VC), forced expiratory volume in one second ($FEV_{1.0}$) and peak flow rate (PFR). The subject of these tests:

	Observed (O)	*Predicted* (P)	O/P
VC	4.0	5.3 litres	76%
$FEV_{1.0}$	2.0	4.4 litres	45%
$FEV_{1.0}$/VC%	50%	83%	56%
PFR	200	645 litres/min	31%

(a) is more likely to have been a man of 25 than a woman of 65

(b) is more likely to be suffering from restrictive than obstructive disease of the lungs

(c) could be suffering from either asthma, chronic bronchitis or emphysema

(d) would be likely to have an arterial carbon dioxide pressure 50% above normal.

197.

(a) T this is called atelectasis; the air is absorbed into the capillaries as it would be in a pneumothorax

(b) F the partial collapse of the lung tends to lower intrapleural pressure

(c) F physiological dead space is space which is ventilated but not perfused. Here the "space" is not ventilated

(d) F the fall of O_2 tension in the collapsed section causes constriction of the local blood vessels by a direct action.

198.

(a) F the reverse is true; this shift tends to increase the oxygen content (ordinate), the curve is shifted upwards as well as leftwards

(b) T a shift to the left is associated with a greater affinity of Hb for O_2 and therefore the tendency to release less O_2 to the tissues

(c) F this would aggravate the already impaired oxygen delivery; fortunately increased 2,3-DPG shifts the curve to the right to increase delivery

(d) T this is related to a fall in 2,3-DPG hence transfusion with such old blood could fail to relieve tissue hypoxia.

199.

(a) T the predicted values are those of a man of 25 (height 70″, 1.8 m). For a woman of 65 (63″, 1.6 m), the $FEV_{1.0}$ of 2.0 litres would be normal

(b) F the relatively severe reduction in $FEV_{1.0}$ and PFR are typical of fairly severe obstructive disease; in restrictive disease $FEV_{1.0}$ and VC are reduced to a similar extent

(c) T these all produce a similar "obstructive" pattern of respiratory function

(d) F respiratory failure is a rare complication of obstructive airways disease.

200. A diver breathing air at a depth of 30 metres under water:
(a) is exposed to a pressure of about 4 times that at the surface

(b) has an increased uptake of nitrogen resulting in unconsciousness from nitrogen narcosis

(c) has a 2–3 fold increase in the oxygen content of the blood

(d) expends less energy on the work of breathing than at the surface for a given oxygen consumption.

201. As anaesthesia deepens:
(a) the respiratory response to carbon dioxide excess (hypercapnia) decreases
(b) the respiratory response to hypoxia is more affected than the response to hypercapnia

(c) a progressive fall in pH is likely to occur
(d) spontaneous ventilation ceases when the peripheral chemoreceptors cease to function.

202. Cyanosis (blue discoloration of skin and mucous membranes):
(a) occurs when the arterial blood contains more than 5 g of carboxyhaemoglobin per 100 ml blood

(b) due to lung disease is typically present in the finger tips but not in the tongue
(c) in patients with interventricular septal defect becomes less likely as right ventricular pressure rises

(d) in the toes but not the fingers is associated with a patent ductus arteriosus where the direction of flow is from pulmonary artery to aorta.

200.
(a) T 1 atmosphere of air+3 of water (1 atmosphere = approximately 10 metres of water = approximately 100 kPa)
(b) F nitrogen does go into solution, mainly into the fatty tissues, but unconsciousness does not usually occur until air is breathed at 12 or more atmospheres
(c) F oxygen content goes up by less than 10% (1–2 ml/100 ml); four times as much oxygen will go into solution in the blood (about 1 ml/100 ml blood) and body fluids but the haemoglobin is almost completely saturated at surface pressures
(d) F the work of breathing is increased because the air is compressed to one-quarter of its surface volume and is much more viscous.

201.
(a) T like most other reflex phenomena in states of CNS depression
(b) F the response to hypoxia is maintained rather better and may account for most of the respiratory drive. Oxygen administration may then remove the main drive to ventilation and lead to CO_2 retention
(c) T due to CO_2 retention
(d) F it ceases when the respiratory centre becomes paralysed by the anaesthetic.

202.
(a) F carboxyhaemoglobin (COHb) is a cherry-pink compound; for cyanosis to be present the blood in small skin vessels must contain at least 5 grams of deoxygenated haemoglobin/100 ml
(b) F it is present in both places (central cyanosis)
(c) F it becomes more likely; the higher the right ventricular pressure, the greater the tendency for venous blood to pass into the left ventricle
(d) T the stream of deoxygenated blood from the ductus passes mainly to the descending aorta.

203. A patient whose lung disease causes CO_2 retention is likely to have:

(a) a raised blood bicarbonate level
(b) an alkaline urine

(c) a lowered arterial blood Po_2 when breathing air at atmospheric pressure
(d) cool extremities.

204. Removal of one lung (e.g. because of lung cancer):

(a) is unlikely to reduce the $FEV_{1.0}$ by more than 10%

(b) leads to cyanosis at rest
(c) results in a halving of the ventilation/perfusion ratio
(d) is likely to cause the heart to shift towards the side of the absent lung.

205. Artificial ventilation for respiratory failure:

(a) may cause tetany if carried out at an excessive rate

(b) requires the use of a humidifier to prevent airway drying if applied through a tracheostomy (opening into the trachea)
(c) is necessary for a patient suffering from spinal cord transection at the level of the second cervical segment
(d) is usually required when the forced expiratory volume in one second has fallen to 50% of normal due to disease of the respiratory muscles or nerves.

206. Coughing:

(a) is reflexly initiated by mechanical or chemical irritation of the alveoli
(b) is depressed during anaesthesia

(c) results in a high expiratory flow velocity due to powerful contractions of the diaphragm
(d) differs from sneezing in that the glottis is closed at the beginning of the expiratory effort.

203.

(a) T to compensate for the respiratory acidosis

(b) F the urine is usually acid since the tubular reabsorption and manufacture of bicarbonate to compensate for the respiratory acidosis is associated with increased hydrogen ion secretion

(c) T impaired ability to excrete CO_2 implies impaired ability to take up the less diffusible oxygen

(d) F they are characteristically warm—due to the direct vasodilator effect of CO_2 on peripheral blood vessels.

204.

(a) F it reduces it by slightly more than half; although the remaining lung may expand, the ability to ventilate it falls slightly

(b) F not if the remaining lung is healthy

(c) F the ratio is little affected

(d) T the mediastinum is drawn towards the relatively empty space.

205.

(a) T the respiratory alkalosis results in increased binding of ionic calcium to the plasma proteins

(b) T the humidifying effect of the upper respiratory tract has been lost

(c) T all the respiratory muscles will be paralysed (phrenic nerve arises from $C_{3,4,5}$)

(d) F this is compatible with adequate ventilation at rest.

206.

(a) F it is initiated by irritation of the trachea and bronchi

(b) T this may allow mucus to accumulate in the bronchial tree and, if obstruction occurs, regional lung collapse will follow

(c) F the high velocities are due to powerful contraction of *expiratory* muscles in the presence of bronchoconstriction

(d) T coughing is more "explosive" than sneezing.

207. In chronic lung disease such as chronic bronchitis and emphysema:

(a) the total ventilation/perfusion ratio is a poor indicator of the severity of the disease
(b) the rate of oxygen uptake at rest is typically reduced

(c) the residual volume is increased

(d) peak expiratory flow rate (PFR) is usually reduced.

208. After filling their lungs completely, three subjects (male) expired as rapidly and completely as possible to produce the records shown below. The subjects were of similar size and age and subject A had no evidence of lung disease. Subject:

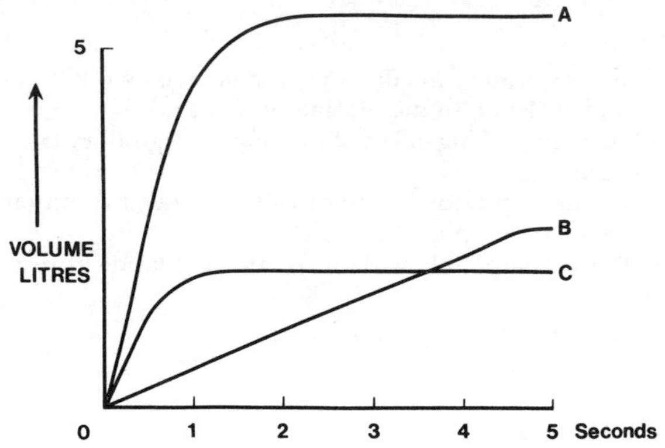

(a) B has an abnormally low ratio |forced expiratory volume in one second | : |vital capacity | ($FEV_{1.0}$/VC)
(b) B is likely to have a higher peak expiratory flow rate (PFR) than subject C
(c) C has a record typical of obstructive airways disease
(d) B is unlikely to survive removal of one lung for lung cancer.

207.

(a) T it may be normal in the presence of severe maldistribution of air and blood flow

(b) F it tends to rise; oxygen uptake in the long term is a function of metabolic rate and the work of breathing is typically increased in chronic lung disease

(c) T due at least partly to closure of small airways at higher-than-normal lung volumes

(d) T it is reduced in proportion to the severity of the disease.

208.

(a) T at about 20% it is extremely low

(b) F as indicated by the slope of the trace, it is likely to be much lower

(c) F it is typical of restrictive lung disease

(d) T his $FEV_{1.0}$ is already dangerously low, i.e. less than 25% of the predicted normal.

209. In obstructive airway disease those patients who are dyspnoeic but not cyanosed ("pink puffers") differ from those who are cyanosed and oedematous ("blue bloaters") in that they have a lower:

(a) forced expiratory volume in one second ($FEV_{1.0}$)
(b) red cell mass (haematocrit . blood volume)
(c) arterial blood pH
(d) sensitivity to carbon dioxide.

210. Patients with restrictive lung disease differ from those with obstructive airways disease by having a lower:

(a) ratio of forced expiratory volume in one second/vital capacity ($FEV_{1.0}/VC$)
(b) peak flow rate (PFR)
(c) residual volume (RV)
(d) total lung capacity (TLC).

211. A region of the lung where the ventilation/perfusion (V/P) ratio is:

(a) low would tend to lower systemic arterial oxygen content
(b) high would have a similar effect on systemic arterial blood gases to a septal defect which allowed blood to pass directly from the right to the left atrium
(c) low would tend to increase the physiological dead space
(d) high would tend to raise systemic arterial carbon dioxide content.

209.

(a) F the two groups have a similarly low $FEV_{1.0}$

(b) T unlike the blue bloaters they tend not to have secondary polycythaemia

(c) F the blue bloaters tend more towards acidosis, due to CO_2 retention

(d) F the lower sensitivity of the blue bloaters could contribute to their poor ventilation, hypoxia and hence pulmonary hypertension.

210.

(a) F the ratio is normal or even raised in restrictive disease, lowered in obstructive disease

(b) F changes in PFR tend to parallel changes in $FEV_{1.0}$/VC

(c) T RV falls in restrictive disease, rises in obstructive disease

(d) T in restrictive disease both VC and RV fall; in obstructive disease TLC changes little (VC decreases, RV increases).

211.

(a) T by lowering alveolar Po_2 and hence the oxygen content of the blood leaving the region

(b) F a low V/P ratio would have a similar effect to such a septal defect (systemic arterial O_2 content reduced, CO_2 content increased)

(c) F dead space is ventilated but not perfused (V/P = infinity)

(d) F it tends to lower the CO_2 content.

ALIMENTARY SYSTEM
BASIC PHYSIOLOGY

212. The diagram below shows the pH of samples obtained through a tube which had been passed into the stomach of a normal person. An injection of either pentagastrin (a synthetic gastrin mimic) or an inert solution was made at the time shown by the arrow. In this situation:

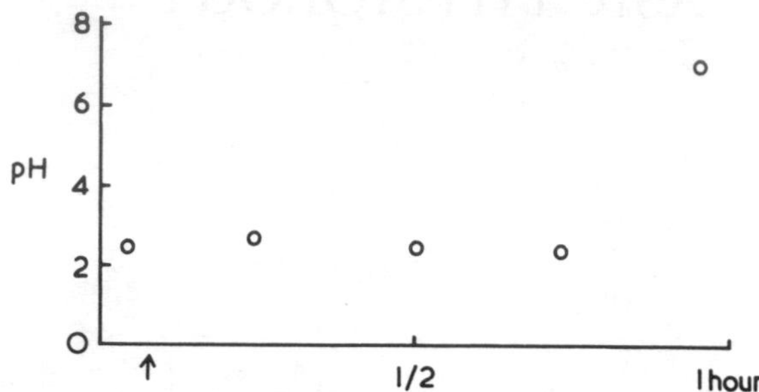

(a) the result for the initial sample confirms that gastric secretions have, in fact, been obtained
(b) the low initial pH suggests that the subject had taken a meal within half an hour of the beginning of the test
(c) the injection is likely to have been inert because it produced little fall in pH
(d) the final sample is more likely to have been obtained from the duodenum than the stomach.

213. Bile:
(a) salts contribute to the solubility of cholesterol in the bile
(b) contains bilirubin in the unconjugated rather than the conjugated form
(c) contributes more than pancreatic secretions to the neutralisation of acid from the stomach
(d) contains a higher concentration of bicarbonate after concentration in the gall-bladder.

212.

(a) T only the stomach would provide fluid with such a low pH

(b) F this is a normal fasting pH

(c) F pentagastrin (and gastrin) increase the volume rather than the acidity of gastric secretion

(d) T the sudden finding of a neutral pH is consistent with the tube having passed into the duodenum by peristalsis.

213.

(a) T they form, in association with phospholipids, a micellar solution

(b) F conjugation and then excretion take place in the liver

(c) F bile contains only about one fifth the bicarbonate concentration of pancreatic secretion (secretin-stimulated)

(d) F the bicarbonate concentration falls to less than half (i.e. less than one tenth that in pancreatic secretion); the concentrations of bile salts, pigments and cholesterol rise several-fold.

214. Saliva:

(a) contains a higher concentration of potassium than does plasma
(b) is more acid when secreted at a high rate than when secreted at a low rate
(c) contains calcium ions in a concentration less than half that of plasma
(d) contains more than twice as much iodide as plasma.

215. Swallowing:

(a) is a reflex action co-ordinated in the cervical and thoracic segments of the spinal cord
(b) is associated with elevation of the larynx which is more effective in preventing the entry of food into the airways than is the occluding action of the epiglottis
(c) cannot take place effectively without the function of intrinsic nerves in the oesophagus
(d) can be voluntarily inhibited when the bolus is in the upper third of the oesophagus.

216. Appetite for food:

(a) may be increased when certain areas in the hypothalamus are destroyed
(b) is reduced if the stomach is distended
(c) is lost after the stomach is removed surgically (gastrectomy)
(d) increases as the blood glucose level falls.

214.
(a) T this is so with alimentary secretions generally
(b) F the reverse is true; copious secretion of alkaline (bicarbonate-rich) saliva in response to an acid stimulus helps to maintain an "oral environment" which prevents the teeth from dissolving
(c) F saliva is saturated with calcium ions at a neutral pH; again necessary to maintain the teeth
(d) T under resting conditions the ratio is about 100, though it may be less than 20 at high rates of flow.

215.
(a) F as with other reflexes involving the upper part of the alimentary tract the centre is in the medulla oblongata
(b) T removal of the epiglottis has little effect on the protective mechanism
(c) T these co-ordinate the peristaltic phase of swallowing
(d) F after its initiation, the swallowing process is outside voluntary control.

216.
(a) T bilateral lesions in appetite decreasing areas (satiety centres) cause hyperphagia (excessive food intake)
(b) T due to afferent impulses from the stomach which affect the activity of the hypothalamic appetite centres
(c) F the emotional drive to eat is not dependent on an intact stomach
(d) T hunger is thought to be related to a reduced glucose uptake by certain cells ("glucostats") in the hypothalamus; the availability of other nutrients also influences appetite.

217. Secretion of saliva is increased:

(a) when touch receptors in the mouth are stimulated

(b) more by sweet than by bitter food

(c) just before vomiting

(d) when the parasympathetic supply of the salivary glands is stimulated.

218. Defaecation:

(a) depends on a stretch reflex whose co-ordinating centre is in the lumbar segments of the spinal cord

(b) is initiated by the passage of intestinal contents into the caecum

(c) is more likely to occur just after a meal than just before it

(d) may involve forced expiration against a closed glottis.

219. Thirst:

(a) may be produced by stimulation of certain regions of the hypothalamus

(b) is produced by a rise in blood osmolality even though blood volume is normal

(c) is produced by a fall in blood volume even though blood tonicity is normal

(d) due to water deprivation continues after water is drunk until blood osmolality is restored to normal.

217.

(a) T food, jagged teeth, the dentist's drill and irritating conditions in the mouth are effective stimuli

(b) F the reverse is the case, e.g. bitter agents given as appetite stimulants increase salivation

(c) T this is part of the vomiting reflex; it helps to buffer the acid gastric contents about to pass through the mouth

(d) T this is the main motor path of the salivary reflex.

218.

(a) F the reflex centre is in sacral segments of the spinal cord; it is normally controlled by supraspinal centres

(b) F it is initiated by stretching of the rectum; ileal peristalsis drives small amounts of chyme into the caecum at frequent intervals

(c) T gastric distension may lead to the desire to defaecate (the "gastrocolic reflex"); the mechanism is obscure

(d) T this manoeuvre (Valsalva's) is usually a part of normal defaecation and may raise intrathoracic and intra-abdominal pressure above 100 mm Hg (13.3 kPa).

219.

(a) T neurones in the hypothalamus are concerned in the emotional drive to drink (thirst)

(b) T presumably due to stimulation of osmoreceptors

(c) T presumably due to stimulation of volume receptors

(d) F thirst is temporarily inhibited before this, presumably by stimulation of receptors in the mouth and stomach.

220. The panels below show some effects on pancreatic secretions in a normal subject of two injections, A and B. The upper panels give volume changes, the middle panels changes in bicarbonate concentration and the lower panels changes in amylase concentration. The time course of the observations was one hour. The results suggest that:

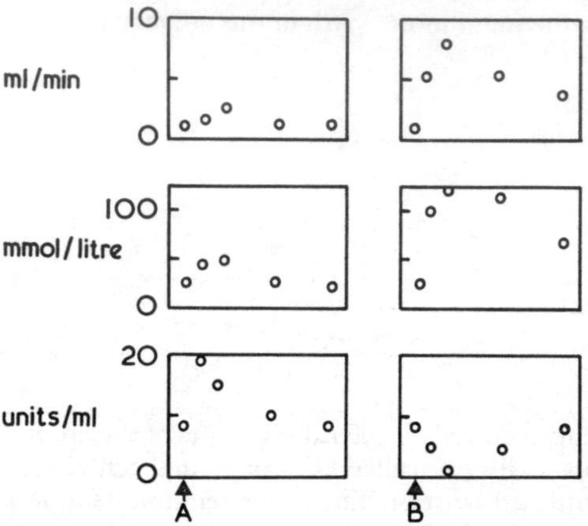

(a) injection A (left panels) and injection B (right panels) were secretin and cholecystokinin-pancreozymin respectively

(b) the initial bicarbonate level in both cases was nearer 200% than 100% of the normal plasma value

(c) at the time of peak bicarbonate concentration in B, bicarbonate ions comprised more than two thirds of the total anions in the secretion, assuming it to be isosmolar with plasma

(d) injection B decreased amylase secretion (units/min) to a minimum less than half the control value.

220.

(a) F injection A produced a small volume rich in enzymes (cholecystokinin-pancreozymin) and B produced a larger volume rich in bicarbonate (secretin)

(b) F at around 25 mmol/litre it was similar to the normal plasma level

(c) T the peak bicarbonate concentration is about 130 mmol/litre; total anions would be around 150 mmol/litre, since the osmolality of plasma is a little under 300 mosmol/litre and most of the particles concerned are univalent ions

(d) F since secretory volume increased by about the same factor as that by which secretory concentration fell, there was no appreciable change in the amount (units/min) of amylase secreted.

221. Gastric secretions:

(a) increase when a hungry person thinks of food

(b) increase in response to the presence of food in the mouth after the vagus nerves to the stomach have been cut

(c) contain a substance which aids the absorption of vitamin B_{12}

(d) cannot digest gastric mucosal cells because the cell membranes contain a pepsin inactivator.

222. Intestinal juice contains:

(a) enzymes which are released mainly in response to vagal activity

(b) enzymes which hydrolyse monosaccharides

(c) the same potassium concentration as extracellular fluid

(d) a substance which activates trypsinogen.

223. Pancreatic secretion:

(a) following vagal stimulation is viscid, scanty and rich in enzymes

(b) stimulated by acid in the duodenum is copious and rich in enzymes

(c) is stimulated in one dog in a cross-circulation experiment when acid is introduced into the duodenum of the second dog

(d) contains enzymes which digest polysaccharides to monosaccharides.

224. The cells of the liver:

(a) are the only important site for synthesis of plasma albumin

(b) are the only important site for synthesis of plasma globulins

(c) can store vitamin B_{12}

(d) release glucose when the blood glucose level tends to fall below normal.

221.

(a) T this is the so-called appetite juice and is a conditioned reflex mediated through the vagus nerves

(b) F the presence of food in the mouth is another stimulus in the reflex ("cephalic") control of gastric secretions; cutting the vagi breaks the reflex arc

(c) T this is called intrinsic factor, a mucoprotein with a molecular weight of 53 000

(d) F they can digest gastric cells but these are normally protected by a coating of mucus.

222.

(a) F the enzymes are contained in mucosal cells; mechanical stimulation is thought to increase the rate of shedding of those cells which then break down and release their enzymes

(b) F there is no evidence for this; monosaccharides are the end-products of carbohydrate digestion

(c) F it is higher; partly because cell turnover rate is high in the intestine and the cast-off cells release their potassium

(d) T this is enterokinase which converts it to trypsin.

223.

(a) T vagal pancreatic juice is secreted reflexly when the subject thinks about or chews food

(b) F acid causes release of secretin which liberates large volumes of alkaline juice which is poor in enzymes

(c) T due to release of secretin in the second dog

(d) F pancreatic amylases digest only as far as the disaccharide stage.

224.

(a) T the regeneration of depleted plasma albumin depends on normal function in the liver

(b) F gamma globulin is produced by plasma cells in the lymphatic system; most other globulins are produced in the liver

(c) T raw liver used to be fed to patients in the treatment of pernicious anaemia (due to vitamin B_{12} deficiency)

(d) T from the breakdown of glycogen.

225. In the large intestine:
(a) the major proportion of intestinal water is absorbed
(b) about one fifth of the intestinal glucose, amino acids and fatty acids are absorbed
(c) mucus-secreting goblet cells are abundant
(d) a substantial proportion of the daily requirement of vitamin A is synthesised by bacteria and absorbed into the body.

226. An increase in body fat:
(a) results in an increase in the percentage of body weight consisting of water
(b) results in a reduction of the lean body mass
(c) increases the specific gravity of the body
(d) decreases survival time in cold water.

227. The respiratory quotient:
(a) is the ratio of the volume of carbon dioxide produced to the volume of oxygen consumed
(b) is 1.0 when glucose is the substrate being metabolised
(c) usually rises in the second week of starvation as energy production becomes increasingly dependent on fat metabolism
(d) tends to fall at the beginning of voluntary hyperventilation.

228. Oxygen consumption is higher than normal:
(a) when the oxygen concentration in the inspired air is higher than normal
(b) when body temperature is higher than normal
(c) when a lightly clad individual is exposed to an environmental temperature of 0°C
(d) during hyperventilation.

225.
(a) F most (several litres) is absorbed by the small intestine; less than a litre is absorbed by the large intestine, converting the faeces from a fluid to a semi-solid state
(b) F these are absorbed in the small intestine
(c) T mucus lubricates and facilitates faecal movements
(d) F the intestinal flora do not synthesise vitamin A; vitamin K is synthesised here and absorbed—vitamin K deficiency may develop when the intestinal bacteria are suppressed.

226.
(a) F the reverse is true since fatty tissue contains little water
(b) F lean body mass = total body weight minus weight of body fat; this stays fairly constant when fat is gained or lost
(c) F it lowers it since fat has a lower specific gravity than the lean body mass
(d) F it favours survival by providing improved skin insulation.

227.
(a) T it depends essentially on the type of food being metabolised
(b) T since $C_6H_{12}O_6 + \mathit{6}\ O_2 \rightarrow \mathit{6}\ CO_2 + 6\ H_2O$
(c) F it falls because the RQ of fat is about 0.7 ($2\ C_{51}H_{98}O_6 + \mathit{145}\ O_2 \rightarrow \mathit{102}\ CO_2 + 98\ H_2O$) and less than that of carbohydrate (1.0) and protein (0.8)
(d) F the reverse is true because CO_2 is washed out at a high rate and O_2 uptake is much less affected.

228.
(a) F O_2 consumption in the long term depends only on the metabolic rate which is not affected by the concentration of oxygen in the air breathed
(b) T a rise in body temperature increases the rate of cell metabolism
(c) T due to the induced shivering
(d) T due to the increased work of breathing.

229. Brown fat:

(a) is relatively more abundant in infants than in adults

(b) is richer in mitochondria than is ordinary adipose tissue

(c) increases its rate of heat production in response to stimulation of its sympathetic nerve supply or increased secretion by the adrenal medulla

(d) is more important than shivering in neonatal thermoregulation.

230. Nitrogen balance:

(a) is neither positive nor negative when the amount of nitrogen excreted by the body equals the amount of nitrogen in the diet

(b) becomes positive when the dietary protein intake is increased

(c) becomes negative when a patient is immobilised in bed

(d) is more negative in late than in early starvation.

231. Saliva:

(a) contains an enzyme which is essential for the complete digestion of starch

(b) is necessary for normal swallowing

(c) keeps the pH in the mouth between 5.0 and 6.0

(d) has an important antiseptic action.

232. The stomach:

(a) is responsible for the absorption of about 10% of the ingested food

(b) depends on the activity of carbonic anhydrase for production of hydrochloric acid

(c) motility increases when fat enters the duodenum

(d) contractions begin in the fundus and travel towards the pylorus.

229.

(a) T it is seen in infants as pads of brown adipose tissue in the neck and interscapular regions

(b) T it has a higher metabolic rate than ordinary fat

(c) T this causes hydrolysis of the triglycerides in the tissue to fatty acids which are metabolised to CO_2 and H_2O in the mitochondria

(d) T infants do not shiver well and the reflex control of metabolism in brown fat is important in maintaining their temperature by "non-shivering thermogenesis".

230.

(a) T this state of *nitrogen balance* implies that there is no net gain or loss of nitrogen

(b) F the additional protein is deaminated and the extra urea excreted; nitrogen balance is maintained

(c) T the immobilisation leads to muscle wasting

(d) T in early starvation body protein is spared to some extent until the carbohydrate and fat stores are exhausted.

231.

(a) F salivary amylase normally initiates the breakdown of starch but is not essential

(b) T on account of its lubricating action

(c) F it maintains mouth pH around 7.0—at more acid levels the teeth start to dissolve

(d) T in its absence, mouth infection quickly develops.

232.

(a) F food products are not absorbed in significant amounts from the stomach

(b) T carbonic anhydrase inhibitors reduce acid secretion

(c) F the reverse is true

(d) T the frequency of the peristaltic waves is about 3/minute; when a wave reaches the pylorus some chyme is squirted through into the duodenum.

233. A gastric pouch:

(a) of the Pavlov type (intact nerve and blood supply) is stimulated to secrete by sham feeding (food taken into the mouth leaves the body by an opening in the oesophagus)
(b) of the Pavlov type produces increased secretions for the same length of time whether the animal takes a normal meal or is sham fed
(c) of the Heidenhain variety (no nerve supply) produces increased secretions for some 3–4 hours after a normal meal
(d) of either type may be stimulated to secrete by injection of secretin into a vein.

234. Stimulation of the duodenum by ingested food leads to hormonal effects which include:

(a) a copious flow of pancreatic juice rich in bicarbonate
(b) decreased gastric motility
(c) contraction of the gall bladder and relaxation of the circular muscle around the lower part of the bile duct (sphincter of Oddi)
(d) a flow of pancreatic juice rich in enzymes.

235. Bile salts:

(a) are the only constituents of bile which are essential for digestion
(b) have a characteristic molecular structure, part water-soluble and part fat-soluble
(c) are to a large extent (more than 75%) reabsorbed from the intestine and resecreted by the liver
(d) are derived from end-products of red cell breakdown.

236. Absorption of:

(a) fats depends on their complete breakdown to fatty acids and glycerol
(b) undigested protein molecules can occur in the newborn
(c) some naturally occurring laevo-rotatory amino acids occurs more rapidly than that of the dextro-rotatory isomer
(d) iron tends to be proportional to body needs.

233.
(a) T by means of a vagal reflex; the response to sham feeding is abolished by cutting the vagus nerves

(b) F only the normal meal causes release of gastrin which leads to more prolonged secretion than the vagal reflex

(c) T this is due to circulating gastrin

(d) F there is some evidence that secretin inhibits gastric secretion of acid.

234.

(a) T this is the action of the hormone secretin
(b) T this has been termed the "enterogastrone" effect
(c) T this is the action of "cholecystokinin" (released when fat enters the duodenum)

(d) T pancreozymin; cholecystokinin (CCK) and pancreozymin (PZ) are believed to be one hormone (CCK-PZ).

235.
(a) T they are essential for emulsification, digestion and absorption of fat
(b) T this characteristic aids fat emulsification by lowering surface tension
(c) T more than 90% are reabsorbed; by this "enterohepatic circulation" the body economises in bile salt production
(d) F this is true of bile pigments; bile salts are derived from cholesterol.

236.
(a) F a proportion of fat is absorbed without being broken down provided it is emulsified into sufficiently small particles
(b) T this occurs by pinocytosis and allows the newborn to absorb the maternal antibodies in the first maternal milk (colostrum)
(c) T because active transport mechanisms involved in their transport can differentiate between the two isomers
(d) T a carrier system in the intestinal mucosal cells has been postulated to account for this; when the carrier system is saturated, the excess iron is excreted in the faeces.

237. The specific dynamic action of food:

(a) is the increase in metabolic rate brought about by eating
(b) is due entirely to the work done in digesting and absorbing the food
(c) persists for about two days after the food is ingested
(d) results in about 30% of the energy value of ingested protein being unavailable for utilisation in the body.

238. Secretion of gastric juice:

(a) is increased when food stimulates cells in the pyloric region of the stomach
(b) is associated with an increase in hydrogen ion concentration of venous blood coming from the stomach
(c) in response to food is reduced if the vagus nerves to the stomach are cut
(d) is essential for protein digestion.

239. In the small intestine:

(a) the concentration of digestive enzymes in the intestinal contents is lower in the ileum than in the jejunum
(b) vitamin B_{12} is absorbed mainly in the jejunum
(c) the final digestion of disaccharides into monosaccharides is carried out by enzymes derived from the intestinal mucosal cells
(d) absorption of the end-products of digestion does not begin until the chyme enters the jejunum.

240. The cells of the liver can:

(a) convert amino acids into glucose
(b) store iron and vitamins A, B_{12}, C and D
(c) inactivate and destroy certain drugs and other compounds in the body by oxidation
(d) inactivate certain hormones by conjugation.

237.
(a) T this is its definition
(b) F it is partly due to the work involved in processing the absorbed material for metabolism or storage
(c) F it usually lasts about 6 hours
(d) T mainly because of the energy required to deaminate amino acids; the figures for ingested fat (4%) and carbohydrate (6%) are much smaller.

238.
(a) T these are the cells which produce gastrin
(b) F bicarbonate ions pass into the circulation as the *alkaline tide*
(c) T vagal activity plays an important role in gastric secretion, particularly of acid
(d) F trypsin and chymotrypsin in pancreatic juice can digest proteins.

239.
(a) T probably because they are digested by proteolytic enzymes (autodigestion)
(b) F it is absorbed mainly in the ileum
(c) T absence of these enzymes interferes with carbohydrate absorption and results in diarrhoea
(d) F though most absorption of foodstuffs occurs in the jejunum, absorption begins in the duodenum.

240.
(a) T this is known as *gluconeogenesis*
(b) F all can be stored in the liver except vitamin C which cannot be stored anywhere in the body
(c) T certain barbiturates and alkaloids such as strychnine are destroyed in this way; in liver failure, the effect of drugs may be potentiated
(d) T progesterone is conjugated with glucuronic acid in the liver to form pregnanediol which is then excreted in the urine.

241. Absorption of dietary fat:
(a) is incomplete if the intestine contains bile salts but not lipase

(b) occurs mainly in the terminal ileum
(c) must occur normally for adequate absorption of vitamins A, D and K
(d) takes place into intestinal lymphatics but not into capillaries.

242. One gram of:
(a) carbohydrate when metabolised by the body yields the same amount of energy as when it is oxidised in a bomb calorimeter
(b) fat when metabolised by the body yields approximately 10% more energy than the metabolism of one gram of carbohydrate
(c) protein when oxidised in a bomb calorimeter yields the same amount of energy as when it is metabolised by the body

(d) protein per kilogram body weight represents an adequate daily protein intake for a person in a sedentary occupation.

243. Plasma cholesterol:
(a) falls dramatically when the intake of cholesterol in the diet is stopped

(b) increases when the amount of saturated fat in the diet is increased

(c) is eliminated from the body mainly by metabolism to carbon dioxide and water in the tissues
(d) is a precursor of adrenal cortical hormones.

244. Free (non-esterified) fatty acids in plasma:
(a) account for less than 10% of the total fatty acids in plasma

(b) are complexed with a protein carrier

(c) decrease when the level of blood adrenaline rises

(d) can be metabolised to CO_2 and water in skeletal and cardiac muscle.

241.
(a) T both lipase and bile salts are needed for normal digestion and absorption
(b) F most is absorbed in the duodenum and jejunum
(c) T these fat-soluble vitamins are absorbed with the fat
(d) F a small proportion of the digested fat (short-chain fatty acids) is absorbed into the intestinal capillary blood.

242.
(a) T the same chemical changes occur
(b) F fat yields approximately 2¼ times as much energy
(c) F it yields about 40% more in the calorimeter; in the body, the urea derived from protein is excreted in the urine and its free energy is not released
(d) T this is probably a high figure; protein requirements increase with increasing levels of energy expenditure.

243.
(a) F plasma cholesterol is derived mainly from synthesis in the body; when the intake falls there is an increase in cholesterol synthesis in the body so that the plasma level is little changed
(b) T cholesterol synthesis in the body is determined mainly by saturated fat intake; substitution of polyunsaturated for saturated fat tends to lower the plasma cholesterol
(c) F cholesterol and its derivatives are excreted mainly in the bile
(d) T its only known function is to act as a precursor of steroid hormones and bile acids.

244.
(a) T the remainder are esterified to glycerol (25%), glycerol+phosphate (50%) and cholesterol (20%)
(b) T they are carried in plasma complexed with plasma proteins
(c) F the catecholamines, adrenaline and noradrenaline, stimulate tissue lipases to release free fatty acids from adipose tissue
(d) T an important source of energy, they can be broken down in most tissues to acetyl CoA which enters the citric acid cycle in the mitochondria.

245. Fat stores in the adult:

(a) normally comprise less than 5% of body weight

(b) release fatty acids in response to adrenergic nerve activity
(c) release fatty acids in response to insulin
(d) increase in size by an increase in adipocyte size rather than by adipocyte multiplication.

246. Metabolic rate can be estimated:

(a) by measuring a person's total heat production over a certain period provided that the energy states of both the person and the measuring device at the end of the period of measurement are the same as at the beginning

(b) from oxygen consumption provided the type of food being metabolised is known

(c) from carbon dioxide production if the respiratory quotient is known
(d) from the type and amount of food consumed in the previous 24 hours.

245.
(a) F up to 15% in men and 25% in women is regarded as normal
(b) T this is a mediated via beta receptors
(c) F insulin favours fat deposition
(d) T the number of adipocytes in the adult remains constant.

246.
(a) T when energy is expended, some appears as work and some as heat because of the inefficiency of the energy transformation. Over a time the energy used for work also appears as heat due to friction, etc. Thus total energy expenditure can be estimated from total heat production
(b) T oxygen consumption in the body is concerned only with the metabolism of food and hence the production of energy. The amount of energy produced by a given oxygen consumption depends on the type of food being metabolised; this can be estimated from the respiratory quotient
(c) T for similar reasons to those mentioned in (b) above
(d) F metabolic rate is not determined by food intake alone; energy intake can be greater or less than energy expenditure in which case the surplus or deficit will be respectively laid down in or removed from the fat deposits.

ALIMENTARY SYSTEM APPLIED PHYSIOLOGY

247. The risk of developing gall-stones tends to:

(a) increase more in iron-deficiency anaemia than in haemolytic anaemia

(b) decrease as the bile salt : cholesterol ratio decreases

(c) increase as the lecithin : cholesterol ratio increases

(d) decrease when supplementary bile salts are taken by mouth.

248. Interference with the normal gastric emptying mechanism:

(a) by cutting all vagal fibres to the stomach, including the pylorus, tends to increase the rate of gastric emptying

(b) by enlarging the pyloric orifice tends to increase the rate of gastric emptying

(c) so that there is unduly rapid gastric emptying may lead to the development of an abnormally high followed by an abnormally low blood glucose level

(d) so that there is unduly rapid gastric emptying may lead to depletion of blood volume and a fall in blood pressure.

249. Poor intestinal absorption of:

(a) iron is a likely consequence of removal of most of the stomach

(b) iodide leads to a reduction in size of the thyroid gland

(c) water occurs in infants who cannot digest disaccharides

(d) calcium is associated with removal of the terminal portion of the ileum.

247.

(a) F increased haemolysis leads to an increased concentration of bilirubin in gall-bladder bile with an increased risk of pigment stones; this is not a feature of iron-deficiency anaemia

(b) F it increases; bile salts are necessary for solubilisation of cholesterol in micelles

(c) F it decreases; lecithin is the main component of the phospholipids which also contribute to micellar formation in bile

(d) T this can be an effective treatment; presumably the ingested bile salts join the bile salt pool via the enterohepatic circulation and improve micellar formation.

248.

(a) F loss of vagal activity tends to reduce seriously the rate of emptying; preservation of fibres to the pylorus in "highly selective vagotomy" tends to avoid this complication

(b) T this may be useful after vagotomy but in some cases excessively rapid passage of food from stomach to intestine may give rise to the "dumping syndrome"

(c) T an increased rate of glucose absorption causes excessive insulin secretion; when glucose absorption is completed abruptly, the insulin remains to cause the low level of glucose

(d) T in this further variety of the "dumping syndrome", the unusually large number of osmotically active particles draw excessive fluid into the gut, thus reducing extracellular fluid and hence blood volume.

249.

(a) T due to loss of gastric acid secretions and rapid intestinal transit; the HCl in gastric juice normally reduces the Fe^{3+} ingested to Fe^{2+}, the form in which it can be absorbed

(b) F. it increases in size (simple goitre); lack of iodide reduces thyroxine production and results in an increase in TSH secretion

(c) T the disaccharides cannot be actively absorbed so they remain in the gut and retain fluid there osmotically; diarrhoea results

(d) F it is associated with removal of the stomach or by pass of the duodenum where its absorption is favoured by a relatively acid medium.

250. Peptic ulcer pain is typically relieved by:

(a) oral administration of substances which neutralise acid, whether or not they are absorbed

(b) oral administration of aluminium hydroxide

(c) a drug which interferes with the action of cholinesterase

(d) a drug which stimulates histamine H_2 receptors.

251. Intestinal obstruction:

(a) is typically associated with constipation

(b) will not cause dehydration if the fluid and electrolytes lost in the vomitus are replaced by intravenous infusion

(c) is associated in the early stages with intermittent vigorous peristalsis

(d) in the colon, but not in the small intestine, gives rise to gaseous distension.

252. A patient suffering from severe diarrhoea characteristically has:

(a) a decrease in the potassium content of the body

(b) a decrease in the sodium content of the body

(c) a raised peripheral resistance

(d) an alkalosis rather than an acidosis.

253. Jaundice is likely to be due to:

(a) hepatic disease if plasma albumin is low and serum bilirubin is mainly unconjugated

(b) common bile duct obstruction if the urine is darker and the faeces paler than normal

(c) haemolytic disease if the prothrombin level in blood is below normal

(d) haemolytic disease if there is a high level of either conjugated or unconjugated bilirubin in the urine.

250.
(a) T these reduce irritation by acid of pain-sensitive nerve endings in the base of the ulcer; non-absorbable antacids are preferable, since, unlike sodium bicarbonate, they do not cause alkalosis
(b) T this is a non-absorbable antacid which, unlike sodium bicarbonate does not carry the risk of systemic alkalosis
(c) F this would increase acid production induced by vagal activity
(d) F this would increase secretion of acid; H_2-blockers reduce secretion of acid and relieve ulcer pain.

251.
(a) T constipation is complete after the bowel below the block has emptied
(b) F dehydration will persist because the large volume of fluid in the distended intestine is also lost to the body
(c) T this causes the intermittent colic typical of early intestinal obstruction
(d) F in both sites stagnation allows bacterial breakdown of cellulose and starch which gives rise to carbon dioxide, methane and hydrogen.

252.
(a) T due to potassium loss in cast-off intestinal cells and in secretions
(b) T intestinal secretions are rich in sodium
(c) T as part of the compensation for a fall in plasma volume and hence arterial pressure
(d) F acidosis is more common due to loss of intestinal bicarbonate.

253.
(a) T since albumin formation and bilirubin conjugation depend on liver activity
(b) T conjugated bilirubin is diverted from faeces to urine
(c) F vitamin K absorption and liver function are likely to be normal in haemolytic disease
(d) F haemolysis causes "acholuric jaundice"; there is a raised level of circulating unconjugated bilirubin but this is not excreted in the urine since it is protein bound.

254. Complications of a partial removal of the stomach (gastrectomy) include:
(a) anaemia

(b) loss of weight

(c) malabsorption of fat due to rapid intestinal transit

(d) severe constipation due to loss of the gastrocolic reflex.

255. Loss of liver function tends to cause:
(a) a fall in the level of alpha and beta but not gamma globulins

(b) increase in breast size in the male

(c) a rise in the level of unconjugated bilirubin in the blood

(d) increased lability of the blood glucose level.

256. Peptic ulceration tends to heal:
(a) following division of the vagal supply of the stomach
(b) after removal of the area of stomach adjoining the pylorus (pyloric antrum)
(c) during treatment with glucocorticoid drugs

(d) during treatment with acetylsalicyclic acid (aspirin).

257. Urobilinogen is:
(a) a mixture of colourless compounds otherwise known as stercobilinogen
(b) formed in the reticulo-endothelial system from bilirubin
(c) converted into the dark pigment urobilin in urine left exposed to the atmosphere
(d) normally absorbed from the intestine and largely excreted in the urine.

254.

(a) T gastric HCl normally helps iron absorption by converting ferric iron to the ferrous form in which it is absorbed; gastric intrinsic factor is needed for B_{12} absorption

(b) T food intake may be seriously reduced due to impairment of the "reservoir" function of the stomach

(c) T due to interference with the delicate mechanism which normally regulates gastric emptying

(d) F the gastrocolic reflex is not essential for defaecation.

255.

(a) T alpha and beta globulins are manufactured in the liver whereas gamma globulins are manufactured by the B lymphocytes

(b) T the increase (gynaecomastia) may be due to the failure of the liver to detoxicate oestrogens

(c) T bilirubin is normally conjugated in the liver with glucuronic acid and excreted in the bile

(d) T by laying down and mobilising glycogen, the liver normally stabilises the blood glucose level.

256.

(a) T this reduces gastric acidity

(b) T gastrin is produced here

(c) F these drugs delay healing by interfering with protein synthesis and mucus secretion

(d) F both glucocorticoids and aspirin impair secretion of the protective gastric mucus.

257.

(a) T it can be found in both urine and faeces.

(b) F its site of formation is the intestine

(c) T hence urine containing much urobilinogen darkens on standing

(d) F it is mainly re-excreted in the bile, an example of enterohepatic circulation.

258. Surgical removal of about 90% of the ileum and jejunum tends to cause:
(a) an increase in the fat content of the stools (steatorrhoea)
(b) demineralisation of the bones (osteomalacia)
(c) a fall in the extracellular fluid volume due to failure to reabsorb sodium chloride and water
(d) anaemia.

259. Loss of exocrine pancreatic function results in:
(a) the presence of undigested meat fibres in the stools
(b) faeces of lower specific gravity than usual
(c) a decreased tendency of the blood to clot, due to fibrinogen deficiency rather than prothrombin deficiency
(d) demineralisation of bone.

260. Liver failure:
(a) is characterised by a low albumin : globulin ratio
(b) tends to reduce the plasma urea level
(c) charactistically causes dependent oedema
(d) may lead to bizarre behaviour in a patient who has just eaten a high-protein meal.

261. In portal hypertension, surgical anastomosis of the hepatic portal vein to the inferior vena cava (portacaval anastomosis):
(a) reduces the patient's risk of developing impaired cerebral function after a high-protein meal (hepatic encephalopathy)
(b) reduces the risk of bleeding into the alimentary tract
(c) increases albumin synthesis by the liver
(d) tends to decrease the amount of ascitic fluid in the peritoneal cavity.

262. Gastric:
(a) acid secretion in response to hypoglycaemia is mediated mainly via the hormone gastrin
(b) emptying is facilitated by its sympathetic nerve supply
(c) acid secretion is increased by administration of histamine together with a histamine H_1 antagonist
(d) acid secretion is inhibited in normal people by histamine H_2 receptor antagonists.

258.

(a) T due to incomplete fat absorption
(b) T calcium, like other nutrients, is trapped in the unabsorbed fat; there is also poor absorption of fat-soluble vitamin D
(c) F adequate salt and water reabsorption can occur in the remaining gut
(d) T due to poor iron and B_{12} absorption.

259.
(a) T there is also an increase in faecal fat
(b) T due to their high fat content (they float in water)
(c) F the reverse; prothrombin deficiency is due to poor absorption of fat-soluble vitamin K
(d) T there is poor absorption of calcium and vitamin D.

260.
(a) T albumin is made in the liver
(b) T urea is synthesised in the liver
(c) T due to plasma albumin deficiency and salt-plus-water retention
(d) T unconjugated toxins derived from protein may produce a state of intoxication.

261.

(a) F the risk is increased; toxins formed by protein degradation passing through the anastomosis bypass the liver
(b) T by diverting blood away from oesophageal varices
(c) F hepatic function is not improved by the procedure
(d) T portal hypertension is an important contributory factor to ascites.

262.
(a) F the response is vagally mediated and is lost after division of the vagal supply to the stomach (vagotomy)
(b) F sympathetic activity delays gastric emptying
(c) T this effect of histamine does not rely on H_1 receptors
(d) T this suggests that histamine plays a role in normal gastric acid secretion.

263. Constipation is a recognised consequence of:

(a) sensory denervation of the rectum
(b) psychological stress
(c) a diet which leaves little unabsorbed residue in the gut
(d) overactivity of the thyroid gland.

264. Vomiting:

(a) expels gastric contents more by skeletal muscle activity than by smooth muscle activity
(b) is co-ordinated at thoracic spinal cord level
(c) is a typical consequence of unilateral removal of a functioning vestibular system (semicircular canals etc.)
(d) is accompanied by autonomic disturbances which can include either a rise or a fall in blood pressure.

265. Obesity is:

(a) unlikely to occur in a person on a high-protein diet even though his food energy intake exceeds his energy expenditure
(b) more likely to cause osteoarthrosis (osteoarthritis) than rheumatoid arthritis
(c) associated with a fall in body specific gravity
(d) not associated with increased mortality until the individual's weight exceeds his "ideal weight" (obtained from tables) by 40% or more.

266. Tone in the lower oesophagus is:

(a) a factor which reduces reflux of gastric contents into the oesophagus
(b) increased by the hormone gastrin
(c) increased by drugs which neutralise gastric acidity (antacids)
(d) increased by anticholinergic drugs.

263.
(a) T this breaks the reflex arc on which defaecation depends
(b) T such stress can modify large intestinal activity to cause either constipation or diarrhoea
(c) T faecal bulk and frequency are directly related to the bulk of food residue
(d) F this leads to increased frequency and fluidity of faeces.

264.
(a) T forced expiratory efforts in the presence of a closed pylorus and relaxed oesophagus are responsible
(b) F it is co-ordinated in the medulla oblongata
(c) T this disturbance of sensory input leads to severe vertigo and vomiting
(d) T a variety of effects on cardiac function and peripheral resistance can either raise or lower the blood pressure.

265.
(a) F if food energy intake exceeds energy expenditure, obesity will occur regardless of the predominant food in the diet
(b) T obesity is a recongised cause of osteoarthrosis in weight-bearing joints
(c) T measuring body specific gravity is a precise but rather impractical way of estimating body fat content
(d) F there is an increase in mortality at +20%; +10% is the conventional level at which obesity is diagnosed.

266.
(a) T this "physiological sphincter" is of considerable importance in preventing reflux and associated pain (heartburn)
(b) T gastrin promotes gastric motility; the increase in oesophageal tone prevents reflux during contractions
(c) T reduction in gastric acidity increases circulating gastrin levels
(d) F these reduce tone, suggesting that cholinergic nerves play a part in maintaining tone.

NEUROMUSCULAR SYSTEM
BASIC PHYSIOLOGY

267. In the diagram below of the neuromuscular junction of skeletal muscle, appropriate labelling would include:

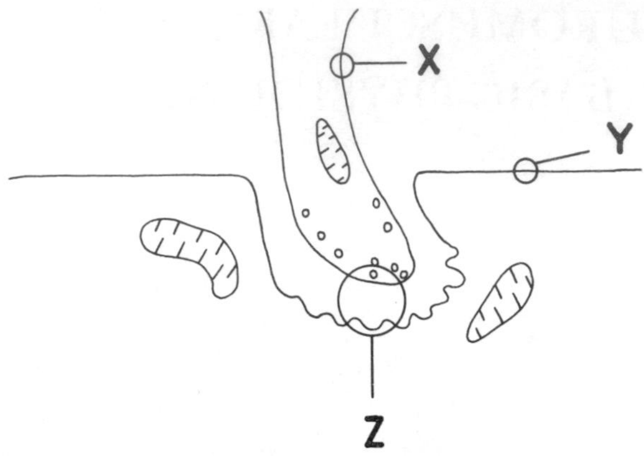

(a) X: part of the motor end plate

(b) Y: the most suitable region for recording miniature (depolarisation) potentials

(c) Z: a region rich in acetylcholinesterase

(d) Y: a region where sodium flux is higher during muscular activity than at rest.

268. A reflex action:

(a) may be carried out by skeletal, smooth and cardiac muscle and by glands

(b) is not influenced by higher centres in the brain

(c) results from activity in at least two central nervous synapses in series

(d) may involve simultaneous contraction of some skeletal muscles and relaxation of others.

267.

(a) F this is part of the motor nerve; the end plate is the part of the muscle fibre adjoining the nerve terminal

(b) F these originate at the muscle end plate membrane in region Z

(c) T this allows for rapid breakdown of acetylcholine

(d) T contraction is initiated by the muscle action potential whose propagation here depends on a transiently high sodium flux.

268.

(a) T e.g., knee jerk; contraction of a full bladder; slowing of the heart and salivation

(b) F higher brain centres influence but do not initiate reflex actions

(c) F the knee jerk reflex arc contains only one central nervous synapse

(d) T in the knee jerk, extensors of the knee contract; flexors relax (reciprocal inhibition).

269. Cerebrospinal fluid:

(a) is a simple ultrafiltrate of plasma

(b) provides for the greater part of the nutritional requirements of the central nervous system

(c) is important in protecting the brain from injury when the head is moved

(d) has a pressure slightly lower than that in the cerebral venous sinuses.

270. A skeletal muscle cell:

(a) has a resting membrane potential such that the inside of the cell is negative with respect to the outside

(b) contains intracellular stores of Ca^{2+}

(c) is normally innervated by more than one motor neurone

(d) becomes less excitable when its resting membrane potential is decreased.

271. In sensory receptors:

(a) stimulus energy is converted into a local potential change which is not propagated but is proportional to the strength of the stimulus

(b) serving touch sensation, constant suprathreshold stimulation causes action potentials to be generated at a constant rate

(c) the frequency of action potentials that they generate may be doubled by doubling the strength of the stimulus

(d) a generator potential can be generated by only one variety of stimulus in any particular sensory receptor.

269.
(a) F its crystalloid pattern differs from that of plasma suggesting that active secretion contributes to its formation
(b) F it has little nutritive function; the blood circulation meets nutritive and gas exchange needs
(c) T by its cushioning and by buoyancy effects; this is its main function
(d) F it is higher; the pressure gradient accounts for its reabsorption by filtration in the arachnoid villi of the dural venous sinuses.

270.
(a) T it is against this background that action potentials are produced
(b) T Ca^{2+} is held in a swollen part of the sarcoplasmic reticulum and is released on excitation to induce an interaction between actin and myosin
(c) F a single motor neurone normally innervates a group of muscle fibres which operate together as a "motor unit"
(d) F this moves the membrane potential towards the threshold potential for excitation.

271.
(a) T this is the *generator potential* which stimulates the adjacent axon to produce impulses
(b) F adaptation is associated with a fall in frequency
(c) F the frequency is related to a logarithm of the strength of the stimulus
(d) F a generator potential may be generated by a variety of stimuli but the receptor is most sensitive to one sort of stimulus (the adequate stimulus for that receptor).

272. The diagram below shows electromyograms (EMGs) recorded at the ankle in response to electrical stimulation applied to the back of the knee. The stimulus in the case of the lower trace was twice that in the case of the upper trace and the latency of response in each case is shown in milliseconds (ms). The recordings are consistent with:

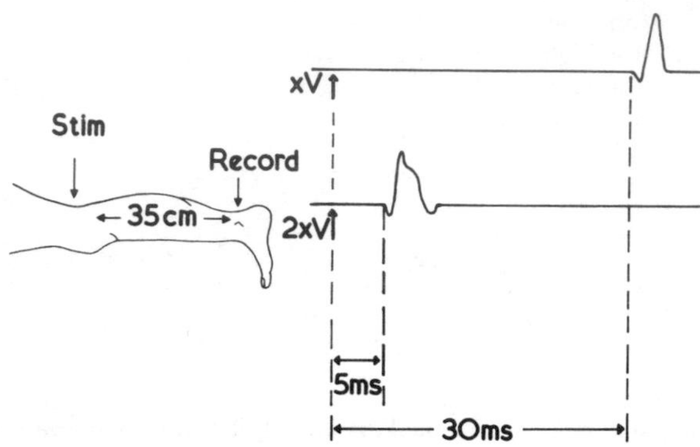

(a) a lower threshold for stimulation of motor nerves than sensory nerves

(b) a rate of conduction in motor fibres nearer 35 than 70 metres per second

(c) a uniform velocity of conduction in all the motor axons stimulated

(d) a synaptic delay at spinal cord level of up to 25 ms.

273. A somatic lower motor neurone:

(a) carries impulses to skeletal muscle but not to other neurones

(b) often innervates more than one muscle fibre

(c) conducts impulses at the same speed as an autonomic lower motor (postganglionic) neurone

(d) may conduct impulses which cause skeletal muscle to relax.

272.

(a) F the reverse is true; the upper EMG resulted from reflex muscle contraction following stimulation of sensory nerves at the knee

(b) F the impulse travels 35 cm in approximately 5 ms, giving a velocity around 70 metres per second

(c) F uniform velocity would give rise to a single narrow peak in the lower EMG

(d) F comparison of the records shows that the impulse took some 25 ms to travel from knee to spinal cord and back; by far the greater part of this time would be taken in traversing the nerves.

273.

(a) F it carries impulses which excite the Renshaw cells in the anterior horn; these inhibit the original neurone thus providing negative feedback

(b) T the coarser the muscle movement the more fibres supplied by one motor neurone. The muscle fibres supplied by a single motor neurone comprise a motor unit

(c) F the somatic neurone (A α fibre) conducts much faster (60–120 metres/second) than the autonomic neurone (C fibre) which conducts at around 1 metre/second

(d) F all somatic motor impulses are excitatory to skeletal muscle.

274. Impulses serving pain sensation in the left foot are relayed:
(a) in the left spinothalamic tract
(b) in the same spinal cord tract which serves heat sensation
(c) to the thalamus on the right side
(d) in the right internal capsule to the precentral gyrus where they enter consciousness.

275. An excitatory postsynaptic potential (EPSP):
(a) is the reversal of the polarity across the postsynaptic nerve cell membrane which occurs when a presynaptic neurone is stimulated
(b) may be recorded from an electrode in a posterior root ganglion cell
(c) is conducted at the same rate as an action potential in a given nerve
(d) is caused directly by the electric field induced by electrical activity in the presynaptic nerve terminals.

276. The ascending reticular formation:
(a) is activated by collateral branches of sensory neurones transmitting impulses to the thalamus
(b) sends impulses to most parts of the cerebral cortex
(c) when stimulated electrically tends to increase alertness
(d) transmits impulses to higher centres via a multisynaptic pathway.

274.
(a) F they cross to the right spinothalamic tract
(b) T pain and temperature fibres go together in the spinothalamic tract
(c) T all sensations other than smell are relayed to the thalamus
(d) F pain enters consciousness at a subcortical level, probably in the thalamus; the precentral gyrus is motor.

275.
(a) F it is a transient (about 5 msec), small (about 5 mV) depolarisation of the membrane towards the threshold for firing
(b) F it may be recorded from a motor neurone in the anterior horn of the spinal cord grey matter
(c) F an EPSP is not conducted
(d) F it is due to release from the presynaptic nerve of a transmitter substance which increases transiently the permeability of the postsynaptic membrane to certain ions.

276.
(a) T an increase in sensory input increases activity in the formation
(b) T the diffuse cortical activity which follows sensory stimulation is abolished by damage to the reticular formation
(c) T the level of alertness parallels the level of activity in the ascending reticular formation
(d) T the diffuse cortical activity which follows sensory stimulation occurs later than the local activity in the postcentral gyrus.

277. In contrast with the awake but resting state, sleep is associated with a lower

(a) metabolic rate

(b) rate of urine formation

(c) arterial level of carbon dioxide
(d) skin temperature in the hands.

278. In the diagram below of autonomic pathways, appropriate labelling would be:

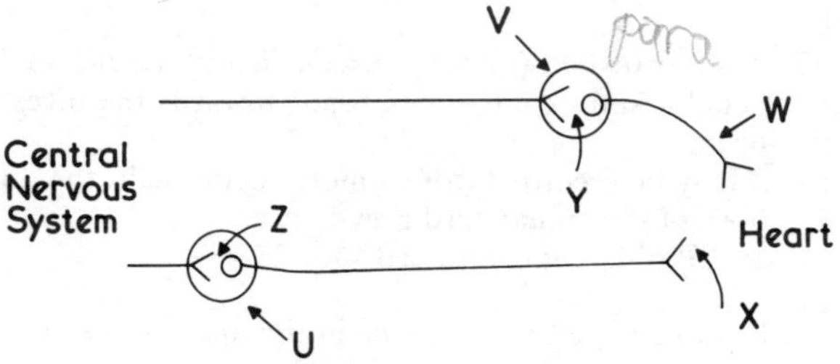

(a) U: parasympathetic ganglion; V: sympathetic ganglion

(b) W: cholinergic transmission
(c) X: transmission by adrenaline

(d) Y and Z: both cholinergic transmission.

279. The cerebellum:
(a) modifies the discharge of spinal motor neurones
(b) is essential for finely co-ordinated movements

(c) has an afferent input from the motor cortex
(d) has an afferent input from proprioceptors.

277.

(a) T at least in part due to a lower blood level of catecholamines

(b) T at least in part due to a higher blood level of antidiuretic hormone

(c) F the level rises due to relative hypoventilation

(d) F skin temperature tends to rise due to vasodilatation.

278.

(a) F the ratio preganglionic axon length : postganglionic axon length indicates the reverse

(b) T acetylcholine is the parasympathetic transmitter

(c) F noradrenaline is the sympathetic transmitter (circulating adrenaline comes from the adrenal medulla)

(d) T both sympathetic and parasympathetic ganglia have cholinergic transmission.

279.

(a) T via the pyramidal and extrapyramidal tracts

(b) T damage to the cerebellum results in movements being clumsy (ataxia)

(c) T this provides information of the desired movement

(d) T this provides information on the actual movement; the cerebellum can thus compare the desired movement with the actual performance and correct for any discrepancy between them.

280. In the region of the motor end-plate of skeletal muscle:
(a) the nerve ending contains many vesicles and mitochondria
(b) are acetylcholine receptors similar to those in smooth muscle

(c) lack of calcium ions diminishes the end-plate potentials in response to nerve stimulation

(d) there is a high concentration of an enzyme cholinesterase which breaks down acetylcholine.

281. The blood-brain barrier:
(a) results in certain molecules in the blood taking longer to equilibrate with tissue fluid in the brain than with tissue fluid elsewhere
(b) permits CO_2 to pass freely

(c) is more permeable to water-soluble than to fat-soluble substances

(d) is more permeable in infants than in adults.

282. A nerve impulse:
(a) can travel in one direction only in any axon

(b) can travel in one direction only across a synapse

(c) is conducted along an axon at approximately the speed of electric current in a conductor
(d) has a duration which corresponds approximately with the absolute refractory period for the nerve.

280.
(a) T the vesicles are thought to contain acetylcholine
(b) F the receptors are thought to be different since they are blocked by different drugs; curare in skeletal and atropine in smooth muscle
(c) T calcium ions seem to be necessary for the release of acetylcholine and the resulting action potentials which lead to contraction
(d) T this is necessary to prevent unduly prolonged muscle contraction.

281.
(a) T this may be a function of the cerebral capillaries or of surrounding tissue
(b) T O_2, CO_2 and nutrients pass easily to meet the metabolic needs of brain cells
(c) F fat-soluble substances such as ethyl alcohol can cross the barrier more readily than water-soluble substances such as creatinine
(d) T the blood-brain barrier is less well developed in newborn animals.

282.
(a) F electrical stimulation of an axon results in two impulses which travel in opposite directions away from the point of stimulation
(b) T from the presynaptic nerve terminals to the postsynaptic dendrites
(c) F it is an entirely different process and much slower
(d) T the axon can not be re-excited when the polarity of the membrane is reversed.

283. The diagram below indicates some sensory pathways and their connexions in the spinal cord. In the case of someone treading on a pin and thereby receiving a sudden painful stimulus to the foot:

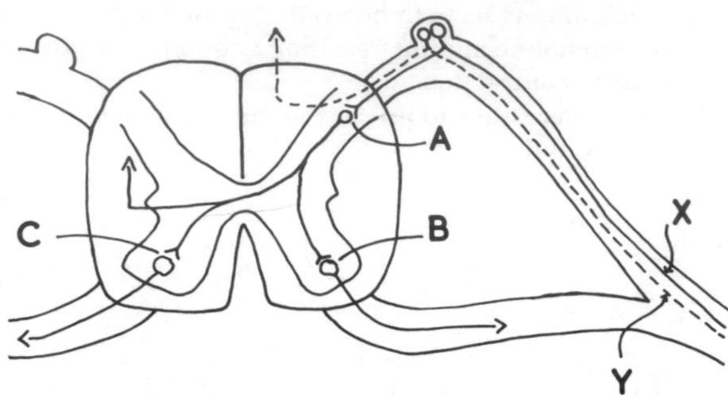

(a) the fibre carrying the sensation of pain would be represented more appropriately by Y than by X

(b) impulses crossing the synapse A would tend to cause depolarisation rather than hyperpolarisation of the postsynaptic membrane

(c) the motor neurone activated at B would tend to supply an extensor rather than a flexor muscle in the leg

(d) the motor neurone activated at C would tend to produce a "crossed extensor reflex".

284. Skeletal muscle:

(a) contains pacemaker cells

(b) contracts when calcium is taken up by the sarcotubular system

(c) contracts when the actin and myosin filaments shorten

(d) contraction strength is related to its initial length.

285. Saltatory conduction:

(a) occurs only in myelinated fibres

(b) velocity decreases as the temperature falls

(c) is faster than non-saltatory conduction in nerve fibres with diameters of 10–20 μm (Aα or Ia type)

(d) velocity is proportional to the diameter of the fibre.

283.

(a) F impulses from Y are carried in the posterior column; impulses from X travel to the spinocerebellar tract which carries pain impulses

(b) T depolarisation (excitatory postsynaptic potential) is necessary for propagation of the impulse

(c) F pain causes withdrawal of the foot (flexion reflex)

(d) T this allows the contralateral leg to support the body while the painful leg is flexed.

284.

(a) F skeletal muscle cells cannot generate spontaneous activity

(b) F *release* of calcium by the sarcotubular system is thought to activate the contraction process

(c) F it contracts when the filaments slide together over one another

(d) T moderate stretch favours contraction.

285.

(a) T a myelin sheath interrupted by nodes of Ranvier is essential; excitation leaps from node to node

(b) T cooling the nerve decreases the rate of sodium conductance at the nodes

(c) T theoretically, saltatory conduction becomes the slower when fibre diameter is less than 1 μm

(d) T and to the distance between nodes.

286. The electroencephalogram (EEG):
(a) provides an indication of a person's intelligence
(b) tends to show waves of smaller amplitude during deep sleep than in the alert state
(c) shows waves with a lower frequency during intense thought than during sleep
(d) is bilaterally symmetrical.

287. Parasympathetic nerves:
(a) generally have opposite effects to those of sympathetic nerves when both supply the same organ
(b) play an important part in the vasodilatation in skeletal muscle during prolonged exercise
(c) liberate acetylcholine at both preganglionic and postganglionic nerve endings
(d) tend to have longer postganglionic than preganglionic fibres.

288. Alpha adrenoceptors:
(a) may be distinguished from beta receptors using the electron microscope
(b) are stimulated by adrenaline and noradrenaline
(c) are involved in the constriction of skin arterioles by adrenaline
(d) are involved in the acceleration of the heart by noradrenaline.

289. Primary neurones serving muscle proprioception:
(a) synapse with secondary neurones whose axons pass up the spinal cord in the posterior (dorsal) column
(b) have their cell bodies in the posterior horn of the spinal cord
(c) synapse with neurones which cross the midline of the body in the brain stem
(d) conduct impulses at a similar rate to the neurones which innervate extrafusal skeletal muscle fibres.

286.
(a) F it provides no indication of intelligence
(b) F the reverse is the case
(c) F low-frequency waves are characteristic of deep sleep
(d) T asymmetry is a sign of disease.

287.
(a) T e.g. intestine, bladder, bronchi, sinoatrial node
(b) F skeletal muscle does not have a parasympathetic nerve supply; local metabolites cause the vasodilatation
(c) T the acetylcholine receptors seem to differ at the two sites
(d) F the reverse is true.

288.
(a) F both receptors are theoretical concepts based on drug antagonism and have not been visualised
(b) T also by some drugs with a similar structure
(c) T skin arterioles are thought to contain mainly alpha receptors
(d) F acceleration of the heart is a beta effect; noradrenaline can stimulate beta as well as alpha receptors.

289.
(a) F the primary neurone axons pass up the posterior column and synapse with secondary neurones in the gracile and cuneate nuclei
(b) F they are in the posterior root ganglia
(c) T the crossing (sensory decussation) takes place in the medulla
(d) T both are large fibres, 10–20 μm diameter (Aα, Ia) which conduct at 60–120 metres/second. Rapid conduction shortens reflex delay.

290. Impulses can be conducted along nerve fibres in a fluid medium:

(a) when the extracellular sodium normally present is replaced by potassium
(b) when the extracellular sodium normally present is replaced by a non-diffusible cation
(c) when the temperature is lowered from 37 to 30°C

(d) for some minutes after the sodium-potassium pump has been blocked by metabolic poisons.

291. The alpha rhythm of the electroencephalogram:

(a) is an electrical potential which can be described in terms of its amplitude and frequency characteristics
(b) has a lower frequency than the delta rhythm

(c) disappears when the subject closes his eyes

(d) indicates that the subject is awake.

292. If the resting membrane potential of a nerve cell falls (i.e. becomes less negative), there is a net:

(a) gain of sodium ions into the cell

(b) loss of potassium ions from the cell

(c) loss of protein ions from the cell
(d) loss of chloride ions from the cell.

293. The primary sensory ending of a muscle spindle in a voluntary muscle is stimulated by:

(a) shortening of an antagonist muscle
(b) relaxation of the muscle when under load
(c) stimulation of the gamma efferent fibres to the spindle

(d) shortening of the extrafusal fibres.

290.

(a) F this would depolarise the fibre completely

(b) F an influx of cations is essential for the depolarisation phase of the action potential

(c) T nerves in human skin often function at temperatures below 30°C, but conduction is slowed

(d) T until the electrochemical gradients generated by the pump have been dissipated.

291.

(a) T the amplitude is about 50 microvolts; frequency is 8–12 Hz

(b) F the upper limit of the delta rhythm is 3.5 Hz

(c) F the alpha rhythm is best seen when the subject's eyes are closed

(d) T the alpha rhythm is replaced during sleep.

292.

(a) T a fall in membrane potential increases the permeability of the membrane to Na^+; the entry of Na^+ causes further depolarisation

(b) T the electrical gradient drawing K^+ into the cell is reduced; the K^+ efflux antagonises the effect of Na^+ influx and tends to stabilise the resting potential

(c) F the membrane is impermeable to protein

(d) F Cl^- tends to enter the cell since the opposing negative charge in the cell is reduced.

293.

(a) T this stretches the agonist muscle and its spindles

(b) T this also stretches the spindles

(c) T this causes the muscle elements in the spindle to contract and stretch the sensory region in the middle

(d) F this reduces the stretch of the spindle.

294. In the diagram below showing a section through the lumbar region of the spinal cord with adjoining anterior and posterior roots, the area:

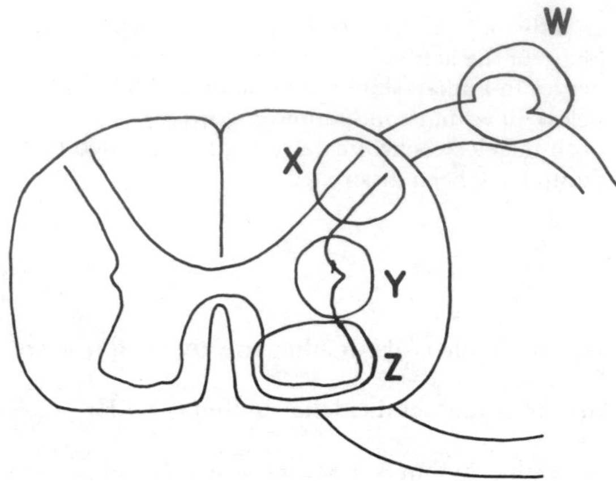

(a) X rather than the area Z would be the site where large fibre afferents could influence the transmission of impulses serving pain sensation
(b) Z rather than the area Y would be the site of cell bodies for sympathetic preganglionic motor axons
(c) X rather than the area W would be the site of the cell bodies of peripheral axons serving pain sensation
(d) Z rather than the area X would be the site of the single central nervous synapse of a reflex arc mediating the knee jerk stretch reflex.

295. The cerebral cortex:
(a) in the occipital region is particularly concerned with hearing
(b) directly concerned with movements of the face and hand has a larger area than that concerned with movements of the legs and trunk
(c) on the left side is more concerned with speech than is that on the right in most people
(d) is darker than the tissue underlying it.

294.

(a) T this is where "gating" could occur, possibly mediated by enkephalins, which are relatively abundant in this region of the cord

(b) F these cell bodies lie in the lateral horn—area Y

(c) F all the peripheral sensory axons, including those serving pain, have their cell bodies in the posterior root ganglion—area W

(d) T the fibre from the muscle spindle synapses with the anterior horn cell motor neurone supplying a motor unit in the quadriceps muscle.

295.

(a) F it is concerned with vision; hearing is appreciated in the temporal cortex

(b) T the area of cortex is related to the precision of the movements concerned rather than to the bulk of the muscles

(c) T generally in right-handed people and sometimes in left-handed people

(d) T hence "grey matter" as opposed to "white matter".

296. A generalised increase in activity in the sympathetic nervous system is characterised by:
(a) contraction of the radial muscle in the iris

(b) increased peristaltic movements of the intestine
(c) a raised blood level of catecholamines

(d) a fall in the blood glucose level.

297. The vagus nerve:
(a) when stimulated has little direct effect on the strength of ventricular contraction
(b) contains afferent and efferent nerve fibres
(c) exerts at rest a tonic effect on the heart which becomes more marked with regular long distance running
(d) contains parasympathetic postganglionic fibres.

298. The action potential:
(a) is initiated by a reduction in the membrane potential to a critical value
(b) is the result of a transient increase in the permeability to sodium and a decrease in the permeability to potassium in the nerve cell membrane
(c) is associated with a reduction in the electrical resistance of the nerve membrane
(d) produced by stimuli of varying intensity is of constant amplitude in a given axon.

299. Non-myelinated axons differ from myelinated in that they:
(a) are more excitable
(b) lack nodes of Ranvier

(c) are not capable of regeneration

(d) are not associated with Schwann cells.

296.

(a) T dilatation of the pupil is due to contraction of the radial dilator muscle
(b) F peristalsis decreases
(c) T partly released by the adrenal medulla, partly by sympathetic nerve endings
(d) F the blood glucose level rises due to breakdown of glycogen in the liver.

297.
(a) T it does not supply ventricular muscle to any important extent
(b) T as do most nerves
(c) T athletes have a slow resting pulse
(d) F it contains parasympathetic preganglionic fibres; the ganglia are in the organs supplied.

298.
(a) T this brings the potential to the threshold value for firing an action potential
(b) F it is the result of a transient increase in sodium permeability followed by a more prolonged increase in the permeability to potassium
(c) T due to increased ionic permeability of the membrane
(d) T the configuration of the action potential is independent of the strength of the stimulus ("all or none" law).

299.
(a) F myelinated fibres have a lower threshold for stimulation
(b) T only myelinated fibres have the nodes; the myelin sheath is interrupted at the nodes
(c) F peripheral myelinated and non-myelinated fibres can regenerate
(d) F both types are associated with Schwann cells.

300. The permeability of the cell membrane of a resting neurone:

(a) to organic anions is greater than to chloride anions

(b) to potassium ions is greater than to chloride ions

(c) to sodium ions is greater than to potassium ions

(d) is such that chloride ions can diffuse through the membrane with the same ease with which they diffuse through water.

301. Acetylcholine:

(a) is thought to act on the same type of receptor in sympathetic and parasympathetic ganglia

(b) is thought to act on the same type of receptors at autonomic postganglionic and somatic motor nerve endings

(c) is broken down in the circulation at a faster rate than noradrenaline

(d) is the transmitter released by some sympathetic postganglionic fibres.

302. Smooth muscle:

(a) is capable of developing tension for long periods of time with relatively little energy expenditure compared with skeletal muscle

(b) when stretched tends to depolarise and contract

(c) contains no myofilaments

(d) contains actin and myosin.

303. An inhibitory postsynaptic potential (IPSP):

(a) may be recorded in a postganglionic neurone in a sympathetic ganglion

(b) drives the membrane potential towards the equilibrium potential for potassium by increasing the permeability of the postsynaptic membrane to potassium and chloride

(c) may result from afferent nerve activity which simultaneously gives rise to EPSPs in other neurones

(d) may summate in space and time with other EPSPs and IPSPs generated on the postsynaptic nerve cell membrane.

300.
(a) F organic anions, which comprise most of the anions within the neurone, do not cross the cell membrane readily
(b) F the permeability to chloride is twice that to potassium
(c) F the permeability to potassium is about 100 times that to sodium
(d) F the permeability of the resting membrane to chloride is several million times less than the permeability of water to chloride.

301.
(a) T its action is blocked at both sites by the same drugs, e.g. hexamethonium
(b) F different drugs are required to produce blockade at the two sites—atropine and curare respectively
(c) T many times faster by cholinesterase in the blood
(d) T e.g. by sympathetic fibres to sweat glands and by sympathetic vasodilator fibres to muscle.

302.
(a) T this is appropriate for its function of maintaining tension for long periods, e.g. in blood vessels
(b) T this "myogenic response" tends to resist stretch
(c) F filaments running in the longitudinal axis of the muscle fibres have been visualised by electron microscopy
(d) T actin and myosin have been extracted; thick and thin filaments are present as in other types of muscle.

303.
(a) F it may be recorded in the cell body of a motor neurone in the anterior horn when certain presynaptic neurones are stimulated. No IPSPs have been recorded in autonomic ganglion cells
(b) T this causes a small (about 5 mV) transient (about 5 msec) hyperpolarisation of the postsynaptic membrane. It is thought to be due to release of transmitter substance (perhaps gamma amino butyric acid)
(c) T impulses from muscle spindles cause simultaneous EPSPs and IPSPs in the motor neurones of agonist and antagonist muscles respectively
(d) T these potentials do summate; the activity of the motor neurone depends on the algebraic sum of the potentials.

304. A volley of impulses travelling in a presynaptic nerve causes:
(a) an identical volley in the postsynaptic nerve
(b) an increase in the permeability of the presynaptic nerve terminals to calcium
(c) vesicles containing preformed transmitter in the nerve ending to fuse with the membrane and release their contents
(d) an increase in heat production by the nerve.

305. Pain receptors:
(a) may be stimulated by potassium ions
(b) may increase their sensitivity in the presence of tissue injury
(c) in the skin are stimulated by all conditions which lead to tissue damage
(d) in internal organs are stimulated by all conditions which lead to tissue damage.

306. The equilibrium potential (E) for any ion:
(a) is the membrane potential that would be required to balance the concentration gradient for that ion
(b) depends on the ratio of the concentrations of that ion inside (I_i) and outside (I_o) the cell
(c) is the potential that the membrane potential would approach if the membrane became freely permeable to that ion
(d) would be zero if the concentrations of that ion on each side of the membrane were equal.

307. A property shared by:
(a) skeletal and cardiac muscle is their striated appearance at microscopy
(b) skeletal and multi-unit smooth muscle is their lack of activity in the absence of nerve stimulation
(c) cardiac and visceral smooth muscle is their spontaneous activity in the absence of nerve stimulation
(d) multi-unit and visceral smooth muscle is that neither has a stable resting transmembrane potential.

304.
(a) F the synapse may amplify or diminish the signal
(b) T when the impulse reaches the nerve terminals calcium ions are taken up by them
(c) T calcium uptake facilitates this process (exocytosis)
(d) T though very small, heat is generated by the nerve impulse.

305.
(a) T this may contribute to the pain experienced when skeletal muscle exercises with a deficient blood supply
(b) T tissue injury lowers the threshold of the pain receptors so that trivial stimuli may give rise to pain (hyperalgesia)
(c) T chemicals released from damaged cells probably cause the stimulation
(d) F internal tissue damage may be painless, e.g. widespread destruction of liver, kidney or adrenal.

306.
(a) T e.g. a membrane potential of about +65 mV would be required to balance the concentration gradient of sodium across the nerve cell membrane
(b) T it can be calculated using the Nernst equation: $E = 61 \log_{10} (I_i)/(I_o)$
(c) T e.g. the neuronal membrane potential would go from its resting value of −90 mV to +65 mV if its membrane were freely permeable to sodium
(d) T from the Nernst equation ($\log_{10} 1 = 0$).

307.
(a) T both have a highly organised arrangement of actin and myosin
(b) T the iris is an example of multi-unit smooth muscle
(c) T isolated hearts and isolated segments of gut show spontaneous activity
(d) F multi-unit smooth muscle, unlike the visceral type, does not contain pacemaker cells so that the membrane potential in individual cells is stable.

308. During sleep as compared with the waking state:

(a) the pupils are more dilated
(b) mean blood pressure tends to fall
(c) the blood level of growth hormone is lower
(d) the blood catecholamine level (adrenaline and noradrenaline) is higher.

309. Histological and physiological study of skeletal muscle shows that:

(a) sarcomere length (between two Z lines) remains constant during muscle contraction
(b) the width of the anisotropic (A) band remains constant during muscle contraction
(c) tension developed is maximal when the actin and myosin molecules just fail to overlap
(d) the stimulus necessary to cause local contraction is minimal when applied at the Z line.

310. In rapid eye movement (REM) sleep as compared with non-REM sleep:

(a) the electroencephalogram (EEG) shows waves of higher frequency
(b) general muscle tone is higher
(c) heart rate and respiration are more regular
(d) dreaming is commoner as judged by recall on awakening at the end of the phase.

308.
(a) F the pupils are constricted during sleep
(b) T heart rate also decreases
(c) F it is higher; growth and cell repair may be favoured by sleep
(d) F it is lower.

309.
(a) F it shortens as the muscle shortens
(b) T its width is the length of the myosin molecule
(c) F it is minimal at this stage; it is maximal when actin and myosin overlap maximally (before adjoining actin molecules overlap)
(d) T this is where the transverse tubules penetrate the muscle fibre.

310.
(a) T amplitude is also smaller, as in the lighter stages of sleep
(b) F despite the EEG appearances, muscle tone is very low
(c) F they are much more irregular
(d) T subjects woken just at the end of the REM phase give vivid accounts of dreams; woken ten minutes after the end of the phase dreaming is not recalled.

NEUROMUSCULAR SYSTEM APPLIED PHYSIOLOGY

311. The diagram below shows a section through the lower cervical region of the spinal cord. Damage in the:

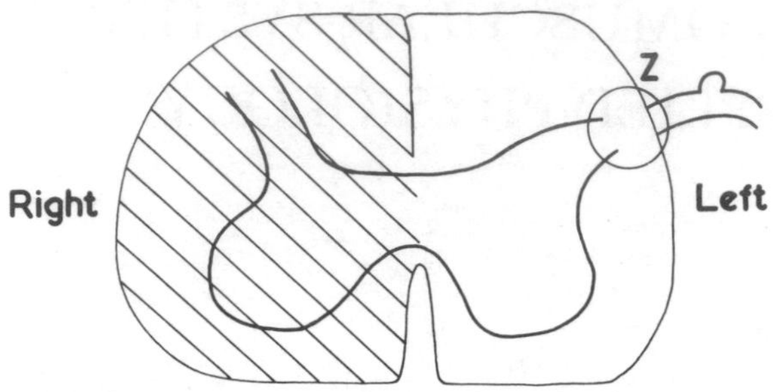

(a) shaded half of the cord would interfere with pain sensation more in the right foot than in the left foot
(b) cord at this level would tend to cause long-term loss of bladder tone rather than an "automatic" bladder
(c) area Z would tend to impair sensation from the left arm while leaving tendon reflexes intact at this level
(d) cord at this level could directly affect the co-ordinating centre for vomiting.

312. Stretch reflexes such as the knee jerk tend to be exaggerated:
(a) when the patient allows his arm muscles to relax
(b) immediately following spinal cord section above the segment responsible for the reflex
(c) on the opposite side of the body some weeks after damage to an internal capsule
(d) in the disorder of the extrapyramidal system known as Parkinsonism.

313. Predominant delta-wave activity in the electroencephalogram:
(a) implies low-frequency activity of very low amplitude as compared with alpha activity
(b) indicates that the patient is thinking deeply
(c) on the left side when there is alpha activity on the right suggests a left-side brain abnormality
(d) is a feature of *petit mal* epilepsy.

311.

(a) F pain fibres from the foot cross the midline soon after entering the cord and well below the cervical region

(b) F the reverse is true; upper motor neurones would be involved, since the reflex centre for micturition is in the sacral cord

(c) F injury to the sensory inflow affects both sensation and the reflex arc

(d) F this centre is situated in the medulla oblongata.

312.

(a) F the knee jerk is more evident (*reinforced*) when the subject puts his arm muscles into a sustained contraction

(b) F immediately following section, there is a state of spinal shock in which all reflex activity below the level of the lesion is lost

(c) T the loss of supraspinal influence eventually results in facilitation of the reflex

(d) F though muscle tone is increased the tendon jerks are not exaggerated in Parkinsonism.

313.

(a) F delta activity is slow but of higher voltage than alpha activity

(b) F it is associated with deep sleep

(c) T delta waves indicate disease or damage as well as deep sleep

(d) F petit mal produces a characteristic "doublet" consisting of a large spike and rounded wave 3 times a second.

314. Increased intracranial pressure:
(a) decreases cerebral blood flow

(b) tends to cause enlargement of the cranium when it occurs in children
(c) tends to produce cupping of the optic disc—the disc bulges backwards away from the vitreous humour

(d) in children may cause squinting and loss of smell sensation.

315. Pain receptors in the hollow viscera may be stimulated by:
(a) cutting through the wall with a sharp scalpel

(b) distension of the viscera
(c) inflammation of the wall

(d) vigorous rhythmic contractions of the smooth muscle in the wall of an obstructed viscus.

316. Signs indicating loss of brain-stem function in a deeply unconscious patient include:
(a) incontinence or urine and faeces
(b) failure of pupils to constrict in response to light

(c) absent tendon jerks in the arms
(d) failure of stimulate ventilation by a gas mixture containing 5% carbon dioxide in oxygen.

317. Excessive vascular constriction in the digits (Raynaud's phenomenon) is more likely to be relieved by:
(a) division of parasympathetic than by division of sympathetic motor neurones
(b) a drug which interferes with transmission at cholinergic nerve endings than by one which interferes with adrenergic transmission
(c) a drug which blocks alpha adrenoceptors than by one which blocks beta receptors
(d) maintaining digital skin temperature at 30°C than by maintaining it at 15°C.

314.

(a) T by compressing cerebral vessels; reduction in blood flow to the medulla may explain the hypertension and bradycardia that often accompany raised intracranial pressure

(b) T because the cranial sutures have not closed

(c) F it produces bulging in the opposite direction, *papilloedema*, due to compression of the veins which drain the retinal circulation and accompany the optic nerves

(d) T cranial deformity may damage a variety of the cranial nerves.

315.

(a) F the intestine may be cut painlessly during operations under local anaesthetic

(b) T stretch is an adequate stimulus for these receptors

(c) T this results in a lowering of the threshold for stimulation of pain receptors (*hyperalgesia*)

(d) T this is the cause of the intermittent pain known as *colic*.

316.

(a) F continence is maintained by higher brain centres

(b) T this reflex has its co-ordinating centre in the midbrain which, together with pons and medulla, constitute the brain stem

(c) F these reflexes are mediated in the spinal cord

(d) T reflex hyperventilation in response to carbon dioxide is mediated in the medulla oblongata—the high oxygen concentration should not interfere with the response but minimises the risk of hypoxic damage during the test.

317.

(a) F sympathetic, but not parasympathetic nerves, supply the digits

(b) F the constrictor fibres are adrenergic

(c) T the vasoconstriction is mediated through alpha receptors

(d) T local temperature can modify the response to sympathetic tone in this way.

318. Muscle tone is:

(a) assessed as the resistance offered by a muscle to passive stretch
(b) reduced by curate-like drugs
(c) altered when the function of the semicircular canals on one side is disturbed by disease
(d) increased after loss of function in the cerebellum.

319. When intracranial pressure is dangerously high:

(a) cerebrospinal fluid (CSF) should be removed by lumbar puncture if signs of respiratory depression occur
(b) drugs which tend to reduce the extracellular fluid volume such as diuretics are a useful palliative
(c) the patient's life may be saved by making an anastomosis between a ventricle and a jugular vein
(d) jugular venous compression may be used to cause venous distension and force CSF out of the cranial cavity.

320. Parkinsonism, a disease of the extrapyramidal motor system, is characterised by:

(a) paralysis of one side of the body
(b) tremor which is worse when a skilled movement is being carried out than at rest
(c) an increase in muscle tone which is maintained throughout the range of passive flexion and extension of a joint
(d) an increase in spontaneous facial movements during conversation.

321. Lower motor neurone disease:

(a) is characterised by loss of voluntary movement but preservation of reflex movement
(b) is a later stage in the development of upper motor neurone disease
(c) of long standing is typically associated with wasting of the affected muscles
(d) of recent onset is associated with involuntary contractions of small fasciculi in the affected muscles.

318.
(a) T this is how it is assessed in clinical practice
(b) T these are used to obtain satisfactory muscle relaxation during surgery
(c) T the conflicting evidence from the two sides may so disturb equilibrium that the patient tends to fall
(d) F the cerebellum helps to maintain normal tone; hypotonia is common in cerebellar disease.

319.
(a) F this may cause a fatal compression of the medulla as it is forced (*coned*) down into the foramen magnum
(b) T by reducing blood pressure and the rate of formation of CSF
(c) T this is an effective treatment
(d) F this would aggravate the condition.

320.
(a) F paralysis is not a feature
(b) F the tremor is worse at rest
(c) T this is the "cogwheel" or "lead pipe" type of rigidity in contrast to the "clasp knife" rigidity of upper motor neurone lesions
(d) F there is poverty of facial movement.

321.
(a) F since the link between CNS and muscle is broken, movement cannot occur either voluntarily or reflexly
(b) F it is independent of upper motor neurone disease
(c) T the cause of this "disuse atrophy" is not known
(d) T this "fasciculation" is due to hypersensitivity to acetylcholine in the recently denervated muscle.

322. A drug which blocks alpha adrenoceptors is likely to cause:

(a) a reduction in sweat production

(b) a fall in blood pressure which is due to slowing of the heart

(c) constriction of the bronchi

(d) a reduction in gastro-intestinal motility.

323. Atropine (which blocks the action of acetylcholine at autonomic postganglionic nerve endings) is liable to cause:

(a) difficulty in reading

(b) diarrhoea

(c) difficulty with micturition

(d) constriction of the bronchi.

324. Bulging of the optic disc into the vitreous humour (papilloedema) is caused by:

(a) raised intraocular pressure (glaucoma)

(b) a rise in intracranial pressure

(c) inflammation of the optic nerve as it enters the eye

(d) very high blood pressure.

325. Pain arising from the viscera:

(a) may cause reflex contraction of nearby skeletal muscle

(b) may cause reflex sweating, vomiting or changes in blood pressure

(c) is always felt in the mid-line

(d) is poorly localised compared with pain arising from superficial tissues.

322.
(a) F sweating is mediated by nerves which release acetylcholine
(b) F it will tend to lower pressure by reducing peripheral resistance
(c) F blockade of beta adrenoceptors tends to cause constriction of the bronchi
(d) F stimulus of adrenergic receptors inhibits intestinal muscle.

323.
(a) T due to paralysis of the ciliary muscles responsible for accommodation
(b) F it is liable to cause constipation by interfering with colonic and rectal activity
(c) T by interfering with the action of the cholinergic motor nerves of the bladder
(d) F by interfering with parasympathetic activity it tends to cause bronchodilatation; it also reduces bronchial secretions and is useful in premedication before anaesthesia.

324.
(a) F the reverse—cupping—is produced by glaucoma
(b) T papilloedema is a very important diagnostic sign of raised intracranial pressure
(c) T inflammation causes the nerve to swell; it enters the eye at the optic disc
(d) T the oedema is associated with generalised retinal damage.

325.
(a) T this is seen especially in the abdomen and results in "guarding"
(b) T stimulation of visceral pain receptors may be associated with a great variety of reflex autonomic effects
(c) F pain arising from lateral viscera (biliary system, ureters) is localised to one side
(d) T this is one of the characteristics of visceral pain.

326. Atropine (which blocks the action of acetylcholine at autonomic postganglionic nerve endings) is liable to cause:

(a) an increase in the resting heart rate

(b) weakness of skeletal muscle

(c) an excessive flow of saliva

(d) overactivity of the small intestine.

327. A patient with loss of function of the posterior columns of the spinal cord will exhibit:

(a) diminished vibration sense
(b) some loss of pain sensation
(c) plantar flexion in response to strong stimulation of the sole of the foot

(d) a clumsy gait which is made worse if he closes his eyes.

328. Sensory disturbance consisting of:

(a) pain, impairment of all sensation and abnormal sensations (paraesthesiae) down the back of one leg suggests spinal cord damage
(b) loss of pain and temperature but not touch sense in the upper limbs suggests damage in the spinal cord

(c) loss of two-point discrimination without loss of touch sensation suggests a lesion in the thalamus
(d) loss of all sensations on the left side of the body suggests a lesion in the right internal capsule.

329. Aphasia is:

(a) a disturbance of speech in the absence of paralysis of the muscles required for speech
(b) said to be sensory if the patient can understand spoken speech but cannot construct his own sentences
(c) more commonly associated with right- than with left-sided damage to the cerebral cortex
(d) said to be motor if the patient cannot find words to express his thoughts.

326.

(a) T by removing the vagal tone which is generally quite marked at rest

(b) F acetylcholine receptors in skeletal muscle are not affected by atropine

(c) F it causes dryness of the mouth by interfering with salivary gland activity

(d) F it inhibits intestinal activity.

327.

(a) T this sensation is carried in the posterior columns

(b) F impulses serving pain travel in the spinothalamic tracts

(c) T this is the normal response and depends on the integrity of the upper motor neurones; the sensory side of the reflex is not carried in the posterior columns

(d) T the patient uses his eyes to compensate for the sensory ataxia due to loss of proprioceptive information.

328.

(a) F pain is uncommon in spinal cord lesions. The symptoms would suggest a sensory root or peripheral nerve lesion

(b) T dissociated anaesthesia is common in syringomyelia which damages the pain and temperature fibres as they cross near the central canal

(c) F it suggests a lesion in the parietal cortex where sensory discriminations are made

(d) F it suggests right-sided damage in the brain stem or thalamus. Temperature and pain sensation are probably appreciated at subcortical level.

329.

(a) T this is a definition

(b) F sensory aphasia implies an inability to understand the meaning of words

(c) F the speech area (Broca's area) is usually on the left side

(d) T in motor aphasia the patient knows what he wants to express but cannot find words to do so.

330. When the sole of the foot is stroked firmly:

(a) the great toe tends to plantarflex in a normal person

(b) the great toe tends to dorsiflex when the foot is affected by a lower motor neurone lesion

(c) the opposite leg may extend in a patient with a long-standing spinal cord transection

(d) the great toe will dorsiflex in a patient with damage to the extrapyramidal pathways influencing the foot.

331. Damage to the cerebral cortex in man may result in:

(a) inability to carry out purposive movements in the absence of severe sensory loss or muscle paralysis

(b) loss of ability to identify an object by its tactile characteristics

(c) a clumsy gait

(d) complete blindness in one eye.

332. In the characteristic hemiplegia following a cerebrovascular accident (stroke):

(a) the affected muscles are unable to contract

(b) movements which involve both sides of the body such as respiratory and spinal movements tend to be spared

(c) skilled movements tend to be preserved more than unskilled movements

(d) vocal movements tend to be affected more than swallowing movements.

333. Characteristic features of cerebellar disease include:

(a) muscular weakness

(b) loss of muscle-joint sensation

(c) difficulty in touching an object precisely with the tip of a finger

(d) involuntary eye movements (*nystagmus*) when fixing the gaze on an object.

330.
(a) T plantar flexion is the normal response to this stimulus and constitutes a negative Babinski response
(b) F the muscles controlling the toe will be completely paralysed in a lower motor neurone lesion; the toe will not move
(c) T this is the crossed extensor reflex
(d) F dorsiflexion (a positive Babinski response) is characteristic of a pyramidal system lesion.

331.
(a) T this condition (*apraxia*) is associated particularly with damage to the cortex of the dominant hemisphere
(b) T this is *asteriognosia* which is associated with parietal cortex damage
(c) T this form of *ataxia* may result from damage to either sensory or motor systems in the cortex
(d) F damage in the visual area of the cortex causes loss of part of the visual field affecting both eyes.

332.
(a) F the affected muscles can not be contracted voluntarily but they can take part in reflex, synergistic and other involuntary movements
(b) T perhaps because they have bilateral cortical representation
(c) F the reverse is true
(d) T they fall into the skilled movements group.

333.
(a) F muscle tone may be reduced but not muscle power
(b) F there is no sensory loss
(c) T this is associated with *intention tremor*; the cerebellum is necessary for precise co-ordination of movements
(d) T due to inco-ordination of ocular muscles.

334. Long-term consequences of transection of the spinal cord in the upper thoracic region include:
(a) paralysis of the bladder muscle and loss of the ability to defaecate
(b) loss of the ability to regulate sympathetic tone in leg blood vessels in response to carotid sinus baroreceptor stimulation
(c) loss of reflexes concerned with erection of the penis and ejaculation of semen
(d) severe flexor spasms initiated by stimulation of the cutaneous nerves in the legs.

335. A drug which blocks beta adrenoceptors is likely to cause:
(a) relaxation of sphincters in the alimentary tract
(b) dilatation of the bronchi
(c) a fall in cardiac output in someone with heart failure
(d) a decreased dilatation by the blood vessels of skeletal muscles in response to circulating adrenaline.

336. Cutaneous pain:
(a) is due to overstimulation of receptors serving other sensory modalities
(b) can be elicited more readily if the tissue has recently been injured
(c) is due to excitation of receptors by pain-producing chemical substances in the injured tissue
(d) shows marked adaptation, i.e. decreases in severity in response to a constant stimulus.

337. Headache can be produced by:
(a) mechanical damage to the parietal cortex
(b) dilatation of intracranial blood vessels
(c) the presence of blood in the cerebrospinal fluid (CSF)
(d) loss of CSF following lumbar puncture.

334.

(a) F both micturition and defaecation can occur reflexly (their reflex centres are in the sacral cord) but are no longer under voluntary control

(b) T sympathetic outflow to the legs is isolated from the integrating centres in the medulla oblongata

(c) F these reflexes may still occur—the co-ordinating centres are in the lumbosacral cord

(d) T these are an exaggeration of the normal flexor withdrawal reflex.

335.

(a) F the excitatory effect of catecholamines (adrenaline and noradrenaline) on sphincters is an alpha effect

(b) F it may cause constriction—by blocking the beta receptors which mediate the dilator effect of catecholamines

(c) T by blocking the sympathetic drive to the heart

(d) T this dilatation is a beta effect.

336.

(a) F painful stimuli are thought to act on specific pain receptors

(b) T the reduction in the pain threshold is called *hyperalgesia*

(c) T certain chemical substances found in most tissues, such as histamine, bradykinin, 5-hydroxytryptamine, potassium etc. can elicit pain

(d) F pain in general is a sensation which shows poor adaptation; this is important to maintain protection of the tissues.

337.

(a) F the brain substance does not contain pain receptors

(b) T agents causing intracranial and extracranial vasodilatation give rise to a throbbing headache thought to be due to stretching of pain receptors in the walls of the vessels

(c) T the headache caused by such meningeal irritation is associated with neck stiffness

(d) T with the loss of the support afforded to the brain by the CSF, the brain tends to sag in the cranium and excite pain-sensitive nerve endings in the meninges.

338. General depression of function in brain neurones, caused, e.g. by acidosis, hypoglycaemia or drugs, tend to lead to:

(a) loss of unskilled movements before loss of skilled movements

(b) restlessness before stupor

(c) a progressive fall in electroencephalogram (EEG) amplitude

(d) loss of function of cells in the cerebral cortex before loss of function of cells in the brain stem.

339. Loss of pain sensation in the:

(a) feet tends to lead to ulceration of the skin of the feet

(b) knee joint tends to lead to damage to the bone ends

(c) leg can be produced by surgical division of the spinocerebellar tracts in the spinal cord

(d) face can be produced by surgical division of the facial nerve.

340. Severe pain may:

(a) lower blood pressure by raising the total peripheral resistance

(b) reduce the heart rate by increasing vagal tone to the sinoatrial node

(c) induce vomiting by acting through a reflex centre in the brain stem

(d) cause sweating by activating cholinergic nerves to the skin.

341. Intracranial pressure tends to rise when:

(a) cerebral venous pressure rises

(b) an individual carries out the Valsalva manoeuvre (forced expiration with the glottis closed)

(c) cerebral blood flow increases

(d) arterial carbon dioxide pressure falls below normal.

338.

(a) F higher critical functions and skills are lost at an early stage, e.g. driving skills following alcohol ingestion

(b) T delirium is an earlier stage than stupor

(c) F amplitude increases initially (delta waves are associated with the lighter stages of coma) and later decreases

(d) T conscious thought and voluntary movements are lost at a much earlier stage than respiratory activity and the brain stem reflexes controlling blood pressure.

339.

(a) T probably because minor injury persists unnoticed and does not evoke the normal protective reflex responses

(b) T again, damage to the joint does not evoke protective spasm of the muscles supporting the joint

(c) F pain sensation is carried in the spinothalamic tracts

(d) F it can be produced by division of the sensory root of the trigeminal nerve.

340.

(a) F it may lower blood pressure by reducing the total peripheral resistance

(b) T this can contribute to a fall in blood pressure in the "vasovagal" syndrome

(c) T this centre is the vomiting centre in the medulla oblongata

(d) T these are the cholinergic sympathetic nerves to sweat glands.

341.

(a) T this raises cerebral venous blood volume and hence raises pressure within the cranial cavity whose capacity is fixed

(b) T this impedes venous return and raises cerebral venous pressure; crying and coughing can raise intracranial pressure in this way

(c) T this also raises the volume of blood within the cranium

(d) F this causes cerebral vasoconstriction and reduces blood volume in the cranium and hence intracranial pressure.

SPECIAL SENSES
BASIC PHYSIOLOGY

342. The fovea:

(a) lies where the visual axis impinges on the retina

(b) contains more blood vessels than other parts of the retina
(c) is a relatively thick part of the retina

(d) lies on the temporal side of the optic disc.

343. Endolymph:

(a) lies within the membranous labyrinth
(b) has a potassium concentration closer to that of intracellular fluid than that of extracellular fluid
(c) is electrically negative with respect to perilymph
(d) inertia is responsible for the stimulation of receptors in the semicircular canals during rotatory acceleration.

344. Olfactory sensory cells:

(a) show adaptation

(b) are epithelial cells which synapse with the underlying olfactory nerves
(c) are of little importance in appreciating the flavour of food after it enters the mouth
(d) relay with secondary neurones which pass to the thalamus.

345. Adaptation of the eyes for vision in poor light:

(a) is usually completed in 2–3 minutes
(b) is entirely due to regeneration of rhodopsin

(c) may be partly accomplished by wearing red goggles before leaving the bright environment
(d) is better for foveal than for more peripheral vision.

346. The basilar membrane of the cochlea vibrates:

(a) throughout most of its length when a low pitched note (e.g. 100 Hz) is detected
(b) at the same frequency as the sound detected
(c) more strongly the higher the frequency detected

(d) more strongly the louder the sound.

342.
(a) T objects in the centre of the field of vision are focused on the fovea
(b) F no major vessels overlie the fovea
(c) F the retina is relatively thin at the fovea; the superficial layers are absent at this point
(d) T the blind spot corresponds with the optic disc.

343.
(a) T the perilymph is outside
(b) T the potassium concentration is very similar to that of intracellular fluid
(c) F surprisingly it is positive; it may be as much as +80 mV
(d) T the canals move relative to the endolymph and the hairs in the ampullae are displaced.

344.
(a) T it is the person who comes into the under-ventilated room who notices the smell!
(b) F they are nerve cells with hairs which project from the surface of the epithelium in the roof of the nasal cavity
(c) F onion cannot be distinguished from turnip in the mouth if the olfactory cells are anaesthetised
(d) F there is no evidence that the thalamus is involved in the olfactory pathway.

345.
(a) F it takes about 20 minutes
(b) F it is also due partly to regeneration of cone pigments and to dilatation of the pupil
(c) T red glasses enable one to see in bright light but the red light does not bleach rhodopsin
(d) F the rods lie outside the fovea; an object in poor light is best seen if its image is formed peripheral to the fovea.

346.
(a) T the lower the frequency the farther the vibration travels up the cochlea
(b) T
(c) F frequency determines only which part of the basilar membrane vibrates and at what rate
(d) T loudness is signalled by the frequency of impulses generated by that part of the membrane affected by the sound.

347. The cones in the eye:

(a) are responsible for colour vision
(b) are more sensitive to light than the rods
(c) are associated with higher visual acuity than the rods
(d) have their function more impaired than have the rods by vitamin A deficiency.

348. Raising the concentration of sodium chloride applied to a taste bud:

(a) decreases the sensitivity of the taste bud receptors to sodium chloride
(b) increases the size of the generator potential recorded from a receptor cell
(c) increases the size of the action potentials recorded in nerves serving the taste buds
(d) may lead to activation of the ascending reticular formation.

349. The diagram below shows how the intensity of the light which can just be perceived (the visual threshold) changes with time in a person who has entered a dark room at time O after having been exposed to bright light for several hours (curve ABC). In this diagram:

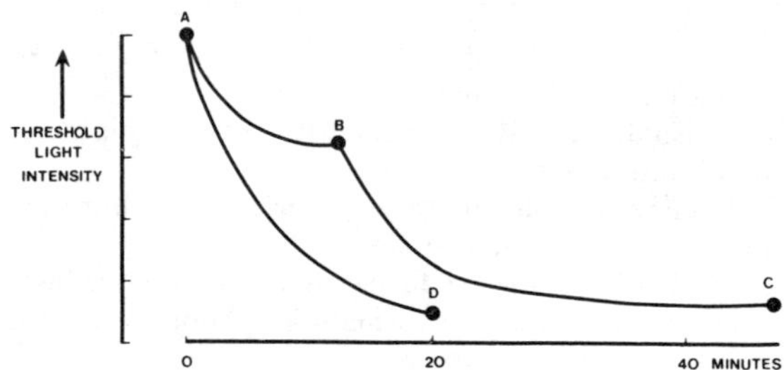

(a) the scale of light intensity would be logarithmic rather than arithmetic

(b) the inflection at B is due to the presence in the retina of two main types of adapting receptor
(c) adaptation from A to B is mainly due to increasing sensitivity of the rods
(d) curve AD is typical of dark adaptation after a much longer light exposure than in the case of curve ABC.

347.
(a) T
(b) F the rods are much more sensitive
(c) T acuity is highest for foveal (cone) vision
(d) F vitamin A is essential only for rod vision (rhodopsin consists partly of vitamin A aldehyde).

348.
(a) T taste receptors adapt to stimulation
(b) T
(c) F there is an increase in frequency, not in size
(d) T all sensory input is thought to activate this system.

349.
(a) T as with other senses, intensity of sensation is related to a logarithm of the stimulus (the threshold light intensity decreases about 10 000-fold between A and C)
(b) T adaptation of the two types of receptor is more or less consecutive
(c) F the earlier and less profound adaptation (AB) is due to increasing sensitivity of the cones
(d) F it is typical of dark adaptation after a much briefer light exposure, e.g. for a few minutes.

350. Constriction of the pupil:
(a) is mediated via sympathetic nerves
(b) increases the refractive power of the eye

(c) increases the depth of focus

(d) decreases the area of retina stimulated by the light.

351. The olfactory system can detect:
(a) odours which are thought all to be combinations of four primary odours, sweet, salty, sour and bitter

(b) a difference in odour between isomers of the same substance

(c) only substances which are volatile and fat- and water-soluble

(d) odours at a lower concentration by sniffing air than by breathing air quietly.

352. During accommodation for near vision:
(a) spherical aberration increases

(b) the ciliary muscle relaxes

(c) the field of vision decreases

(d) the amount of light entering each eye increases.

353. Visual acuity is:
(a) a measure of the sensitivity of the retina to light

(b) better using one eye than using both eyes
(c) better with central than with peripheral vision
(d) reduced in colour-blind persons.

350.
(a) F parasympathetic nerves supply the circular muscle
(b) F it does not affect it but reduces spherical and chromatic aberration
(c) T as in a camera, the smaller the aperture, the greater the depth of focus
(d) F narrowing the aperture of a camera does not result in a smaller photograph.

351.
(a) F these are tastes, not smells; the primary odours are thought to be camphoraceous, musky, floral, pepperminty, ether-like, pungent and putrid
(b) T current theories suggest that molecular shape is the main factor in olfactory discrimination
(c) T otherwise they will not normally reach the olfactory mucosa and be taken up by the receptor membrane
(d) T sniffing directs nasal air currents so that they flow directly over the olfactory mucosa.

352.
(a) F it decreases due to pupillary constriction (most aberration occurs at the periphery of the lens)
(b) F the mainly circular muscle must contract in order to relax the lens ligaments and allow the lens to assume a more spherical shape
(c) T due to convergence of the eyes; there is a greater overlap of the fields of vision perceived by the two eyes
(d) F it decreases due to pupillary constriction.

353.
(a) F it is a measure of the ability of the eye to distinguish between (resolve) two points; it is usually measured using Snellen's letter charts
(b) F the two images reinforce each other
(c) T visual acuity is maximal in the fovea
(d) F it is normal.

354. The tympanic membrane:
(a) modifies the frequency of sound waves impinging on the ear
(b) tends to bulge outwards when the pharyngotympanic (Eustachian) tube is obstructed
(c) stops vibrating almost immediately after the sound stops.
(d) and ossicles lose at least 75% of their effectiveness when a small perforation is present in the membrane.

355. In the refracting system of the eye:
(a) the cornea causes more refraction than the lens
(b) more refraction occurs at the inner surface of the cornea than at its outer surface
(c) the lens can double the refractive power of the eye during accommodation in a young adult
(d) the back surface of the lens contributes more to accommodation than the front.

356. The semicircular canals contain hair cells which are stimulated by:
(a) movement of perilymph
(b) linear acceleration
(c) rotation at constant velocity
(d) cessation of rotation.

357. When light is directed into one eye:
(a) the pupil on that side constricts and the opposite pupil dilates
(b) the pupil can change in size even though the optic nerve has been cut
(c) the pupillary responses are mediated by sympathetic nerves
(d) the pupil does not change in size if the effects of acetylcholine at autonomic nerve endings have been blocked by atropine.

354.
(a) F it faithfully reproduces the frequency of the incoming sounds
(b) F the air in the middle ear is gradually absorbed if the tube is obstructed
(c) T it is very nearly critically damped
(d) F a small perforation has little effect (less than 5 decibels hearing loss); complete loss of the membrane and ossicles causes 50 decibels loss.

355.
(a) T because it has an interface with air
(b) F very little refraction occurs at the inner surface; the refractive indices of cornea and aqueous humour are very similar
(c) F the refractive power of the eye is about 60 dioptres in a young adult at rest; ciliary muscle activity can increase it by about 10 dioptres only
(d) F the images seen when a candle is held in front of the eye indicate that the front of the lens bulges more in accommodation than the back.

356.
(a) F perilymph lies outside the membranous labyrinth; the inertia of endolymph causes it to move relative to the canal wall and displace the hairs in the ampulla when the head is moved appropriately
(b) F they are sensitive to rotatory movements
(c) F at constant velocity the endolymph and hair receptors rotate at the same speed
(d) T deceleration and acceleration are equally effective stimuli.

357.
(a) F they both constrict
(b) F the optic nerve is an essential part of the light reflex pathway
(c) F the pupillary reflexes are normal after section of the sympathetic nerves
(d) T parasympathetic fibres running with the 3rd cranial nerve are responsible for the reflex constriction.

358. Impulses in the auditory nerve:

(a) are thought to be initiated by electrical changes associated with distortion of the hair cells in the cochlea
(b) provide information about the frequency, intensity and waveform of the applied sound
(c) travel via the thalamus to the sensory cortex in the parietal lobes
(d) can lead to reflex contraction of skeletal muscle.

359. In the diagram below the line VXYW represents the threshold of hearing at various frequencies for a normal subject. The:

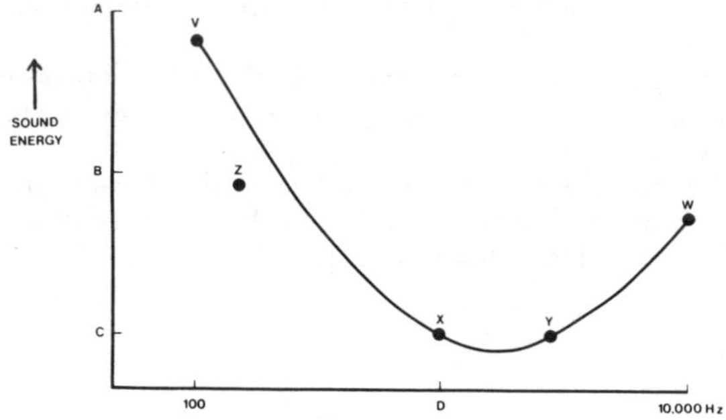

(a) sound waves with the characteristics represented by point Z would be audible to the subject
(b) interval AB on the ordinate would represent 2.0 rather than 20 decibels
(c) point D on the abscissa corresponds to 5 000 rather than 1 000 Hz
(d) segment XY corresponds to the frequencies most important in the auditory perception of speech.

360. Light rays from an object to the right of the visual axis:

(a) are detected in both eyes by the retina to the left of the fovea
(b) generate impulses which are conducted along the right optic tract
(c) generate impulses which give rise to conscious sensation in the eye areas of the frontal lobes
(d) form an inverted image on the retina.

358.
(a) T such electrical changes can be detected as cochlear microphonics
(b) T in the brain this information is interpreted as the "pitch", "loudness" and "quality" of the applied sound
(c) F the cortical area serving auditory sensation lies in the upper part of the temporal cortex
(d) T the stapedius and tensor tympani muscles reflexly damp ossicular movements at high sound intensities.

359.

(a) F anything below the line VXYW is inaudible
(b) F AB and BC both represent 20 decibels
(c) F it corresponds to 1 000 Hz; the frequency (or pitch) scale is logarithmic
(d) T the range 1 000–3 000 Hz (XY) is that to which the ear is most sensitive.

360.
(a) T
(b) F they pass along the left optic tract
(c) F they enter consciousness in the left occipital cortex; the frontal eye fields are concerned with the integration of eye movements
(d) T as in a camera.

361. The tympanic membranes:

(a) tend to bulge inwards during descent in an aeroplane whose cabin is not pressurised

(b) have an area about twice that of the oval windows

(c) when they vibrate, result in vibrations of the oval windows which exert more force per unit area than the tympanic membrane vibrations

(d) increase hearing acuity by preventing the round window from receiving sound waves at the same time as the oval window.

362. The utricle:

(a) is a gravity receptor

(b) communicates with the semicircular canals and the cochlea

(c) contains calcified granules (otoliths) embedded in the matrix surrounding the hairs of the receptor cells

(d) can initiate reflex changes in muscle tone.

363. The rods in the retina:

(a) contain visual pigment which is more sensitive to light in the red than in the blue frequency bands

(b) are rendered insensitive by bright light

(c) are identical with the cones except for the pigments they contain

(d) comprise about one-fifth of the receptor cells in the fovea.

364. Cones:

(a) are found in the layer of the retina closest to the vitreous humour

(b) are confined to the fovea and the retinal area immediately adjacent to it

(c) contain pigments which are less light-sensitive than rhodopsin

(d) are most sensitive to yellow-green light.

361.
(a) T cabin pressure rises above middle ear pressure (swallowing opens the pharyngotympanic tube and equalises pressure)
(b) F their area is about 20 times as great
(c) T the energy gathered by the larger tympanic membrane is concentrated on the smaller oval window
(d) T in the absence of this "round-window protection" vibrations in the inner ear would be greatly damped.

362.
(a) T it signals the strength and direction of the gravitational field; it also responds to linear acceleration
(b) T they are all parts of the membranous labyrinth, filled with endolymph
(c) T the hair cells are distorted in a direction determined by the direction of the gravitational pull
(d) T muscle tone is redistributed so that the body can withstand gravitational stresses.

363.
(a) F they contain rhodopsin which shows maximal light absorption at wavelengths of about 500 nm, the blue-green part of the light spectrum
(b) T in ordinary daylight, nearly all rhodopsin is broken down, i.e. bleached
(c) F they are histologically distinct
(d) F there are no rods in the fovea.

364.
(a) F they are found in the layer of the retina furthest from the vitreous humour
(b) F though the cone concentration is maximal in the foveal region, a small number of cones is distributed over all the light-sensitive parts of the retina
(c) T rhodopsin is the most sensitive of the pigments
(d) T absorption of light by cone pigments is greatest at these wavelengths (550–570 nm).

365. The tympanic membrane and auditory ossicles:

(a) provide impedance matching between the sound waves in air and the vibrations in the cochlea which is nearer 10% than 50% perfect for sound waves around 1 000 Hz

(b) transmit sound with a frequency in the 500–5 000 Hz range with the least loss of energy

(c) are the only means of conducting sound vibrations across the middle ear

(d) transmit sounds less efficiently when the sound amplitude is very great.

366. The receptor cells serving taste:

(a) are primary sensory neurones

(b) are confined to the tongue

(c) initiate impulses which are carried by branches of the trigeminal nerve

(d) lie under the epithelium of the oral cavity.

367. Sound waves:

(a) with an intensity of 0 decibels cannot be heard

(b) with an intensity of 100 decibels do not usually cause pain

(c) sound twice as loud when their intensity is doubled

(d) give rise to a note an octave higher when their frequency increases eight-fold.

368. The frequency of impulses generated by receptors in the utricle:

(a) is related to the orientation of the head with respect to the gravitational field

(b) is higher when the subject travels at 60 than at 30 miles per hour

(c) on one side is inversely related to the frequency being generated in the opposite utricle

(d) is directly related to the frequency of impulses generated by the receptors in the semicircular canals.

365.

(a) F at this frequency, the impedance matching is over 50% perfect; thus most of the energy of the sound wave is transmitted

(b) T impedance matching is optimal and auditory acuity is greatest in this frequency band which corresponds with that of normal speech

(c) F the air of the middle ear can transmit sound waves

(d) T loud sounds cause a protective reflex contraction of the tensor tympani and stapedius muscles which damp ossicular movement.

366.

(a) F they are receptor cells; the terminals of the primary neurones are wrapped around them

(b) F they are found also in the soft palate, pharynx and larynx

(c) F the impulses from the tongue travel in the chorda tympani (anterior two-thirds) and glossopharyngeal (posterior one-third) nerves; from elsewhere they travel in the vagus

(d) F taste bud cells lie in the epithelium, their terminal hair processes project through apical pores in the epithelium.

367.

(a) F by definition a sound of 0 decibels intensity at 1 000 Hz is just above the threshold of normal hearing

(b) T 120 decibels is about the threshold for pain

(c) F there is a logarithmic relationship between sound intensity and loudness; hence the logarithmic decibel scale

(d) F they sound one octave higher when their frequency is doubled.

368.

(a) T the orientation of the head with respect to gravity determines the force applied to the hair cells by the otoliths

(b) F utricle receptors are affected by acceleration, *not* velocity

(c) F often both respond in the same way; it depends on the direction of acceleration

(d) F they respond to different things, the utricle to linear acceleration, the canals to angular acceleration.

369. In the diagram below, the light absorption of the three types of cone receptor at various wavelengths is shown as a percentage of the maximum for each receptor. The wavelengths of blue, green and red light are about 450–500, 500–575 and 650–725 nm (nanometres) respectively. It can be stated that the:

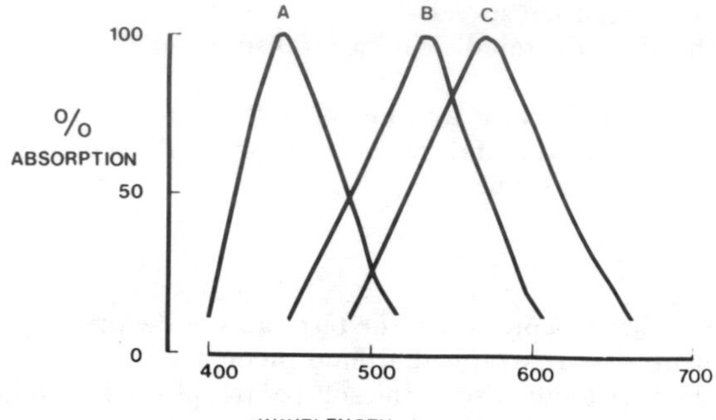

(a) pigment of the A cones reflects red light more completely than does that of the C cones
(b) A cones are more adapted to detection of blue monochromatic light of wavelength 460 nm than are the C cones
(c) C cones are more sensitive to yellow than to red light
(d) curve for the rods would correspond more closely with curve C than with A or B.

370. Aqueous humour is:
(a) formed by active secretion rather than by passive filtration
(b) thought to be formed behind the iris
(c) absorbed into the canal of Schlemm at the corneo-scleral junction
(d) more easily absorbed when the pupil is widely dilated.

369.

(a) T absorption of red light by the A cones is minimal, so reflection is maximal

(b) T they absorb nearly maximally at this wavelength (at which the C cones show little absorption)

(c) T yellow light, between red and green in the spectrum, corresponds to the peak sensitivity of the "red" cones

(d) F its peak lies between A and B; rhodopsin, a red pigment shows minimal absorption of red light.

370.

(a) T the crystalloid composition of aqueous differs from that of plasma and its rate of formation is reduced by enzyme poisons

(b) T the ciliary glands on the anterior surface of the ciliary body are usually given the credit

(c) T blockage of this canal can cause a rise in intra-ocular pressure (glaucoma)

(d) F the iris then tends to close off the entrances to the canal of Schlemm in the corneo-scleral angle.

371. Rhodopsin (visual purple) is:

(a) a purple pigment
(b) most sensitive to violet light

(c) regenerated when the eyes are closed
(d) least sensitive to red light.

372. Below is shown the visual field of a normal left eye as plotted by perimetry. When the eye is focused on point Y, an object at point:

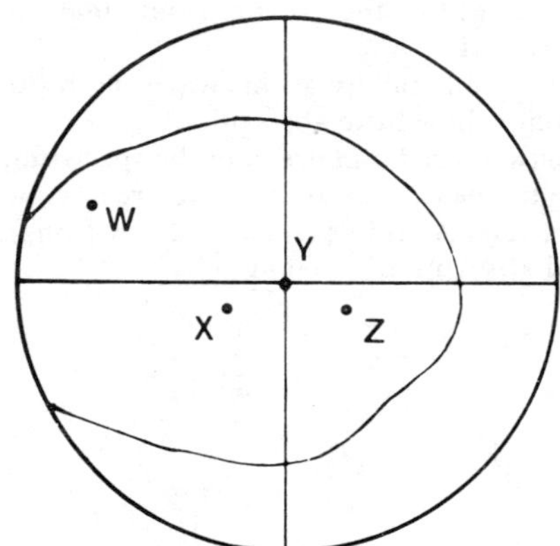

(a) W will be detected in the lower nasal quadrant of the left retina

(b) Y will be detected in the region of the fovea of the macula
(c) Z rather than at point X may be invisible

(d) W will be appreciated as a result of impulses transmitted in the left rather than the right optic tract.

371.
(a) F it is red
(b) F it is most sensitive to blue-green light (wavelength about 500 nm)
(c) T rhodopsin is regenerated in the dark
(d) T the reason the pigment is red is that it reflects (does not absorb) red light.

372.
(a) T the image is inverted and reversed with respect to the object
(b) T the point focused upon is detected at the macula
(c) F the reverse is the case; the optic disc is medial to the fovea, hence the blind spot is in the temporal part of the field of vision
(d) F impulses related to the temporal region of the field of vision cross the midline at the optic chiasma.

373. The basilar membrane:

(a) is broader at the base of the cochlea than at the apex
(b) vibrates with the same frequency as the applied sound

(c) vibrations result in the generation of impulses in the auditory nerve at the frequency of the applied sounds throughout the audible range

(d) of the basal turn of the cochlea is more affected by high-frequency vibrations than is the basilar membrane of the apical region.

374. Taste receptors:

(a) and olfactory receptors are generally more excited by food when it is hot than when it is cold

(b) for sourness are predominant on the tip of the tongue

(c) generally give rise to a salty taste when stimulated by salts of metals
(d) generally give rise to a sour taste when stimulated by hydrogen ions.

373.
(a) F the reverse is true
(b) T the apparatus of the middle ear transmits the applied sound from air to the fluid medium of the inner ear without changing the frequency
(c) F because of their refractory period, nerves can only transmit impulses at frequencies up to about 1 000 Hz. The ear can detect sounds with frequencies of up to 20 000 Hz
(d) T high-frequency sounds travel only a short distance up the basilar membrane (low-frequency sounds travel to the apical region and produce maximal effect beyond the basal turn).

374.
(a) T the receptors are less sensitive at low temperatures; also, fewer volatile molecules reach the olfactory mucosa (unpleasant medicine is less disagreeable when chilled; an exception is quinine which tastes more bitter when cold)
(b) F sweet sensation is experienced at the tip, bitter sensation is predominant at the back, salt and sour at the sides
(c) F sodium chloride tastes salty but lead salts taste sweet and potassium iodide tastes bitter
(d) T all acids taste sour.

SPECIAL SENSES
APPLIED PHYSIOLOGY

375. Four audiograms are shown below, with the upper left diagram representing a normal person's hearing. Open circles represent air conduction and closed circles bone conduction. The:

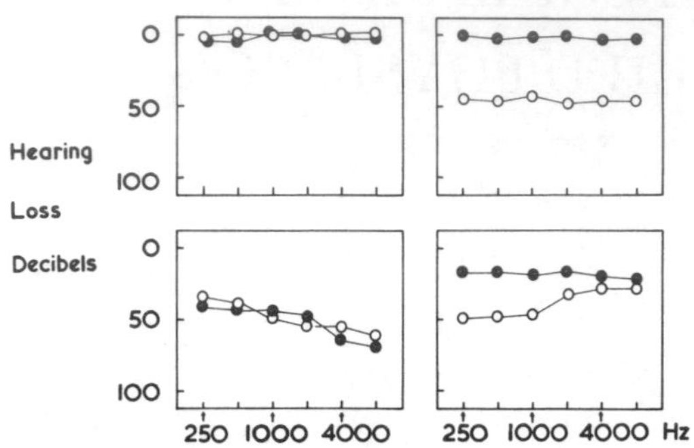

(a) highest frequency marked corresponds to 8 000 Hz rather than 6 000 Hz
(b) upper right diagram represents the effect of ageing rather than stiffness of the auditory ossicles
(c) lower left diagram is consistent with damage to the upper rather than the lower regions of the cochlea
(d) lower right audiogram shows a sensorineural deafness affecting mainly the lower frequencies.

376. Difficulty in maintaining normal balance is more likely:
(a) when there is semicircular-canal damage than when there is cochlear damage
(b) when there is inadequate circulation through the basilar rather than through the carotid arteries
(c) when vision is impaired, in the case of a patient with impaired proprioception
(d) within days rather than within weeks of the surgical removal of one labyrinth.

377. In an individual with short-sightedness (myopia):
(a) the eye tends to be longer than average from lens to retina
(b) use of an appropriate biconvex lens corrects the defect
(c) close vision is more seriously affected than distance vision
(d) a circular object tends to appear oval.

375.

(a) T in the logarithmic scale used, each division represents twice the frequency of the previous division

(b) F the reverse is true; ossicular disorders cause conductive deafness as in this diagram, while ageing causes a sensorineural type of deafness

(c) F this diagram shows maximal hearing loss for the higher frequencies which are detected in the lower regions of the cochlea

(d) F there is a conductive deafness (impaired air conduction) affecting mainly the lower frequencies (e.g. damaged ear drum).

376.

(a) T the canals rather than the cochleae contribute sensory information used in maintaining balance

(b) T the basilar artery supplies brain-stem areas particularly concerned with balance

(c) T vision can compensate to some extent for the impaired proprioception

(d) T the patient's nervous system gradually adjusts to the abrupt cessation of impulses from the absent labyrinth.

377.

(a) T hence distant rays are focused in front of the retina

(b) F this would make matters worse; a biconcave lens is required

(c) F the reverse is true

(d) F this is caused by an asymmetrical cornea (astigmatism).

378. Colour blindness:

(a) is the result of an inability to detect one or more of the three primary colours, red, green and blue

(b) in which the only abnormality is that red and green light cannot be distinguished results from failure to detect either red light or green light but not both

(c) in which there is no sense of colour ("black-grey-white vision") is due to complete loss of cone function

(d) is much commoner in women than in men.

379. Deafness:

(a) for high tones more than for low tones is a typical result of working for years in a very noisy environment (e.g. more than 85 decibels)

(b) in which bone conduction is more effective than air conduction suggests nerve damage on the affected side

(c) due to nerve damage is, in general, more improved by a hearing aid than is obstructive deafness

(d) is often associated with paralysis of the vocal cords (deafmutism).

380. Atropine blocks the action of acetylcholine at autonomic postganglionic nerve endings. Its local application to the eye causes:

(a) dilatation of the pupil

(b) impaired ability to focus on nearby objects

(c) difficulty in looking upwards

(d) a tendency towards impairment of fluid drainage from the anterior chamber.

381. In disease affecting the middle ear:

(a) destruction of the auditory ossicles virtually abolishes hearing

(b) loss of function of the auditory muscles (stapedius and tensor tympani) increases the apparent loudness of sounds of high intensity

(c) obstruction of the pharyngotympanic tube may lead to eardrum rupture when the external pressure rises

(d) immobilisation of the stapes footplate in the oval window produces more severe deafness than removal of the ossicles.

378.
(a) T on the basis of the Young-Helmholz theory, one or more of the three types of cone fails to function
(b) T either the "red cones" or the "green cones" are defective
(c) F it is due to the presence of only one type of functioning cone
(d) F it is 20 times as common in men and is due to recessive characteristics carried by genes on the X chromosome.

379.
(a) T an audiogram showing this is important evidence when industrial compensation is being claimed (basal cochlea damaged)
(b) F this is typical of conductive deafness—usually due to wax or middle-ear disease
(c) F the reverse is true—amplification tends to counteract the loss in the outer or middle ear but cannot make the nerve more efficient
(d) F deaf-mutes are mute because, being deaf, they lack the sensory feedback required for speech.

380.
(a) T by paralysis of the cholinergic constrictor fibres
(b) T the ciliary muscle has a cholinergic innervation
(c) F atropine has no effect on the external ocular muscles.
(d) T by dilating the pupil and causing obstruction by the iris of the corneo-scleral angle.

381.
(a) F sound can still be transmitted across the air of the middle ear; loss is only moderate
(b) T these muscles tend to damp vibration; in their absence noises are unnaturally loud
(c) T middle-ear pressure cannot be equalised to atmospheric pressure (hence it is inadvisable to practise skin-diving when suffering from pharyngitis)
(d) T when the stapes, which is attached to the oval window membrane, is immobilised, the oval window membrane cannot vibrate; deafness may be severe.

382. A typical result of severe damage to the visual pathway at the level of the:

(a) left optic tract is an inability to see objects in the right half of the normal visual field (right homonymous hemianopia)
(b) optic chiasma by a pituitary tumour is an inability to see objects in the nasal half of each visual field (binasal hemianopia)
(c) optic radiation is loss of the visual field on the opposite side to the damage
(d) occipital cortex is loss of foveal vision with preservation of peripheral vision.

383. Squint (strabismus) may result from:

(a) poor vision in one eye in childhood
(b) central suppression of vision in one eye
(c) damage to the cerebellum
(d) damage to the internal capsule.

384. Impairment of the sense of smell:

(a) may be confined to certain odours only, leaving the ability to detect other odours completely intact
(b) generally accompanies the ageing process
(c) typically follows thalamic damage
(d) is a typical early sign of an occipital lobe tumour.

382.

(a) T fibres from the left half of each retina are concerned with vision to the right and pass to the left optic tract

(b) F the crossing fibres are affected; these come from the nasal half of the retina and mediate temporal vision; there is bitemporal hemianopia

(c) T as with damage to the optic tract

(d) F the reverse is true because the fovea is bilaterally represented in the cortex.

383.

(a) T this results in poor fixation of that eye

(b) F this is a consequence not a cause of squint

(c) F this may cause involuntary oscillatory movements (nystagmus) but not squint

(d) T this may cause a paralytic strabismus due to damage to the oculomotor tracts.

384.

(a) T impairment of smell sensation can be selective, e.g. a percentage of the population is unable to smell the freesia flower

(b) T as with the other special senses

(c) F smell pathways do not pass through the thalamus

(d) F it may indicate a frontal lobe tumour.

385. The diagrammatic audiogram below shows hearing loss over a range of frequencies. In this audiogram:

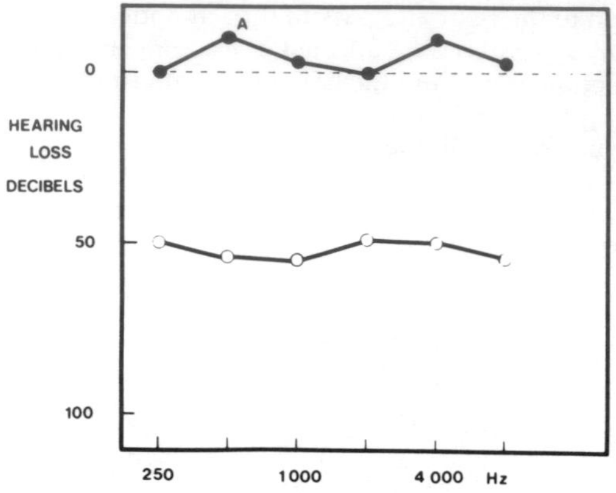

(a) the open circles could indicate results for bone conduction and the solid circles results for air conduction in the same ear
(b) point A indicates better than average hearing at 500 Hz

(c) the features are those of a sensorineural as opposed to a conductive deafness
(d) the features are typical of the deafness due to prolonged exposure to loud noise (traumatic deafness).

386. Involuntary oscillatory movements of the eye (nystagmus):
(a) are not seen in people whose nervous systems are normal

(b) may result from disease of the cochlea

(c) are a feature of cerebellar disease
(d) do not affect the acuity of vision.

385.

(a) F the reverse is true; bone conduction may be unimpaired when air conduction is impaired, but not *vice versa*

(b) T point A indicates that the subject's threshold for hearing is 10 decibels below the average normal threshold at 500 Hz

(c) F impaired air conduction with normal bone conduction is typical of conductive deafness

(d) F traumatic deafness is a sensorineural deafness with maximal loss at around 4 000 Hz.

386.

(a) F they are seen in normal people after a period of rotation on a revolving chair and while watching scenery from a moving train

(b) F they may result from disease of the semicircular canals; sensory information from them is used to regulate eye fixation

(c) T nystagmus in an ataxia of eye fixation

(d) F the rapid eye movements may make vision blurred.

387. When the function of the semicircular canals on one side of the head is impaired:

(a) the patient may experience the sensation that the world is revolving around him
(b) water, just above or below body temperature, introduced into the external auditory meatus on that side causes more prolonged nystagmus than normal
(c) the patient's increased tendency to fall is typically more apparent in the dark than in the light
(d) the patient may experience spontaneous nausea and vomiting.

388. Typical effects of ageing on the special senses include:

(a) movement of the near point towards the eye
(b) a loss of accommodative power of the lens which is nearer 10-fold than 5-fold from infancy to 70 years
(c) loss of hearing which affects bone and air conduction about equally
(d) loss of hearing which affects low- and high-pitched sounds about equally.

389. The diagram shows the path of light falling on the fovea of each eye of a child of three when he looks at a point X. The child:

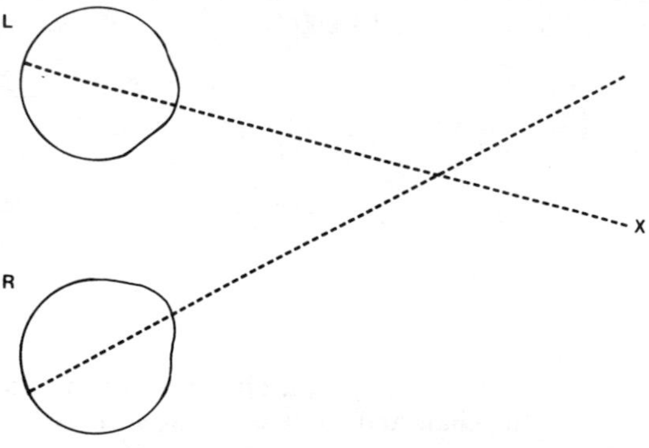

(a) shows a divergent strabismus (squint)
(b) is likely to suppress vision in the left rather than the right eye, assuming that the strabismus is constant
(c) is less likely to suppress vision in one eye if the left eye is covered for part of each day
(d) is more likely than average to suffer from a refractive error.

387.

(a) T the unbalanced information from the two sides may cause this sensation (vertigo)

(b) F the nystagmus is less marked than normal (in this "caloric" test, convection currents in the canals mimic rotation and trigger the eye movements)

(c) T compensating visual stimuli are eliminated in the dark

(d) T unbalanced or excessive stimulation of the semicircular canals, e.g. sea-sickness, can produce a wide variety of autonomic effects as well as nausea and vomiting.

388.

(a) F recession is typical of the ageing eye (presbyopia)

(b) T accommodation is around 10–15 dioptres in early childhood and falls to 5–10 dioptres at 30 and about 1 dioptre at 70

(c) T it is a sensorineural deafness (presbycusis)

(d) F high-pitched sounds are more affected.

389.

(a) F the right eye is converging

(b) F suppression of vision tends to occur in the non-fixing eye

(c) T covering the "good" eye helps to preserve vision in the non-fixing eye

(d) T refractive errors are a cause of squinting.

390. Impairment of visual acuity in bright light can be explained by:
- (a) random light scattering when there is deficient pigmentation of the eye due to albinism
- (b) random light scattering when there is asymmetrical corneal curvature due to astigmatism
- (c) inability to alter the focal length of the lens when a cataract is present
- (d) impairment of rod function when there is vitamin A deficiency.

391. In long-sightedness (hypermetropia):
- (a) objects at infinity cannot be brought to a sharp focus on the retina
- (b) the ciliary muscle must be more strongly contracted than normal to bring an object in the middle of the visual range (say 2–3 metres distant) into clear focus
- (c) the range of unblurred vision (near point to far point) is greater than normal
- (d) the near point can be brought closer to the eye by the use of a biconcave lens.

390.

(a) T normally absorption of light by dark pigment in the choroid prevents back-scattering of light into the retina

(b) F there is a refractive error but not random light scattering

(c) F impairment of acuity in this case is due to random light scattering by opaque material in the lens

(d) F although lack of vitamin A causes night blindness, rod function does not determine visual acuity in bright light; lack of vitamin A can impair acuity due to random light scattering by keratin which is laid down in the corneal epithelium (xerophthalmia).

391.

(a) F this is true of short-sightedness (myopia)

(b) T this distance is closer to the hypermetropic's near point (maximal ciliary action) than to the normal person's near point

(c) F it is less than normal—as in the normal the range extends to infinity, but it starts further from the eye

(d) F a convex lens is required to augment the power of the eye's refracting system.

URINARY SYSTEM
BASIC PHYSIOLOGY

392. Capillary pressure in the renal glomeruli:
(a) is lower than pressure in the efferent arteriole

(b) rises when the afferent arterioles constrict

(c) is higher than in most other capillaries in the body

(d) falls by about 10% when arterial pressure falls by 10% from the normal level.

393. Reabsorption of a filterable substance by the renal tubules is likely to be active rather than passive if:
(a) its concentration in tubular fluid is lower than that in peritubular capillary blood

(b) its excretion is increased by cooling the kidney
(c) its renal clearance value is lower than that of inulin

(d) the ratio (urinary excretion rate) : (plasma concentration) for the substance is the same as for glucose.

394. The renal clearance of:
(a) a substance is expressed in units of volume per unit time

(b) urea is lower than that of inulin

(c) chloride increases after an injection of aldosterone

(d) PAH eventually falls as the plasma concentration of PAH rises.

395. The fluid in the distal part of the proximal convoluted tubule has:
(a) a higher urea concentration than the fluid in Bowman's capsule
(b) a low pH when the kidneys are excreting an acid urine

(c) a glucose concentration similar to that in plasma

(d) an osmolality about $^1/_5$ that in glomerular filtrate due to the active reabsorption of about $^4/_5$ of the filtered sodium chloride in the proximal tubule.

392.
(a) F it is higher; otherwise blood would not flow out of the glomerular capillaries!
(b) F it falls due to the greater pressure drop across the afferent arterioles
(c) T the afferent arterioles offer little resistance compared with other systemic arterioles
(d) F filtration pressure is maintained by a redistribution of renal vascular resistance.

393.
(a) T this would suggest that the substance had been transported into the blood against its concentration gradient
(b) T cooling typically impairs active metabolic processes
(c) F this indicates that there is reabsorption but not whether the reabsorption is active or passive
(d) T anything which is filtered by the glomeruli and has a zero clearance must be actively reabsorbed.

394.
(a) T it is the volume of plasma completely cleared of the substance by the kidneys per minute (ml/min)
(b) T about half of the urea that is filtered is reabsorbed; all the filtered inulin is excreted
(c) F aldosterone increases chloride reabsorption and reduces chloride clearance
(d) T because the secretory process for PAH becomes saturated at high plasma PAH values.

395.
(a) T due to water reabsorption
(b) F acidification takes place mainly in the distal convoluted tubule
(c) F glucose is completely reabsorbed in its passage along the proximal convoluted tubule
(d) F the osmolality is little changed because water is passively reabsorbed in roughly the same proportion as the sodium chloride.

396. Renal tubules normally reabsorb:

(a) a volume of water every hour greater than plasma volume

(b) all the bicarbonate that is filtered when the kidneys are producing an acid urine

(c) more chloride than potassium per unit time

(d) amino acids and plasma protein.

397. In the diagram below, illustrating the handling of glucose by the kidney:

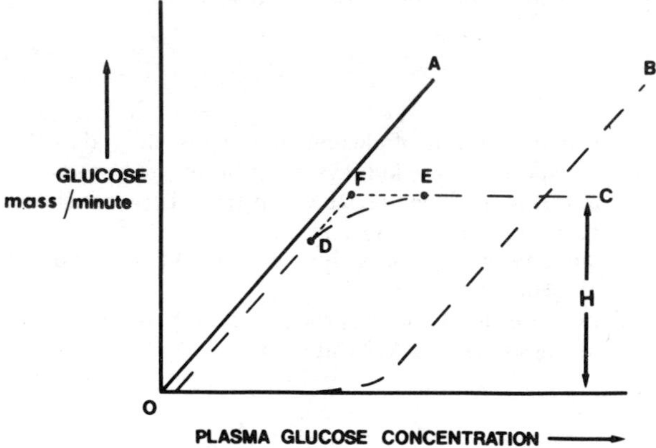

(a) line A could represent the rate of filtration of glucose by the glomeruli

(b) line B could represent the rate of absorption of glucose by the proximal convoluted tubules

(c) the fact that line C follows the curve DE rather than the angle DFE is related to the fact that not all nephrons have exactly the same threshold

(d) H could indicate the maximal reabsorbing capacity of the kidney for glucose.

396.

(a) T the volume filtered per hour is the GFR×60 = about 8 litres. Only about 1 % of this is excreted

(b) T in addition they manufacture bicarbonate to maintain the plasma bicarbonate level at about 25 mmol/l

(c) T the amount of chloride in the filtrate is about 20 times greater than the potassium and the renal clearance of K^+ is greater than the clearance of Cl^-

(d) T though amino acids are freely filtered at the glomeruli, they are absent from the urine; the small amount of plasma albumin which is filtered is also reabsorbed by an active transport system.

397.

(a) T the rate of filtration is directly proportional to the plasma concentration

(b) F line B represents the rate of excretion of glucose; line C represents glucose reabsorbed

(c) T the "splay" is also related to the fact that glucose reabsorption tends to increase slightly in efficiency around the threshold level

(d) T the diagram illustrates features typical of a transport system with a limited maximal reabsorbing capacity.

398. It can be concluded that renal tubular cells secrete a substance into the tubular lumen if:

(a) the clearance value for the substance is over 300 ml/min

(b) it is present in arterial blood and its concentration in renal venous blood is zero

(c) the amount of the substance in the urine is increased by poisoning tubular enzyme systems

(d) its concentration rises as it passes along the proximal convoluted tubule.

399. In the nephron:

(a) fluid in the tip of the loop of Henle is hypertonic with respect to glomerular filtrate

(b) glomerular filtrate is hypertonic with respect to the fluid in the distal convoluted tubule

(c) antidiuretic hormone (ADH) causes the fluid in the collecting ducts to become hypertonic with respect to that in the proximal convoluted tubule

(d) the fluid at the end of the proximal convoluted tubule is hypertonic with respect to glomerular filtrate.

400. Tm- (Transport maximum-) limited reabsorption implies that:

(a) the reabsorption is active

(b) the amount of reabsorption depends critically on the length of time the substance is present in the tubules

(c) below a threshold tubular load, the substance is completely reabsorbed

(d) the renal clearance of the substance so reabsorbed gradually falls as its plasma concentration rises until its Tm- is reached.

398.

(a) T any value above the glomerular filtration rate (about 120 ml/min), indicates secretion

(b) T in this case, secretion is necessary to explain the removal of the substance from the plasma which was not filtered at the glomeruli

(c) F this would suggest that the substance was being reabsorbed by an active process

(d) F this could be explained by water reabsorption, as in the case of urea.

399.

(a) T because the countercurrent system tends to concentrate sodium etc. in the inner renal medulla

(b) T absorption of sodium chloride in the distal part of the loop of Henle and the distal convoluted tubule, where the cell membranes are impermeable to water, results in hypotonicity

(c) T ADH, by increasing the permeability of the collecting ducts to water, allows water to pass out into the hypertonic interstitium of the renal medulla

(d) F if anything, the fluid would be hypotonic, since absorption of water passively *follows* absorption of solutes, however there is little change in tonicity along the proximal convoluted tubule since the walls are permeable to water.

400.

(a) T it is one of the varieties of active tubular reabsorption

(b) F this would be true for *gradient-time limited* varieties of active reabsorption

(c) T any excess above the threshold is excreted in the urine, e.g. glucose

(d) F with Tm limited reabsorption renal clearance remains at zero as the plasma concentration rises until the Tm is reached, when the clearance gradually increases.

401. When a patient's mean arterial blood pressure falls:
- (a) by 50%, renal blood flow falls by less than 10% due to autoregulation
- (b) by 50%, glomerular filtration rate falls by about 50%
- (c) to a level which just abolishes glomerular filtration, the glomerular capillary pressure must equal the intracapsular pressure
- (d) to a level which just abolishes glomerular filtration, renal blood flow will also be abolished.

402. The glomerular:
- (a) capillary pressure is normally similar to that in other systemic capillaries
- (b) afferent arterioles offer more resistance to blood flow than the efferent arterioles
- (c) capillaries are more permeable to water than are most other capillaries in the body
- (d) capillary blood sodium : potassium ratio is the same as that found in glomerular filtrate.

403. If, during an infusion of para-aminohippuric acid (PAH), peripheral venous plasma PAH level is 0.02 mg/ml, urinary level 16 mg/ml and urine production rate 1 ml/min:
- (a) the concentration of PAH in renal venous blood must be higher than 0.02 mg/ml
- (b) the concentration of PAH in renal arterial blood must be about 0.02 mg/ml
- (c) renal plasma flow is nearer 800 ml/min than 1000 ml/min, assuming that the transport maximum for PAH has not been exceeded
- (d) renal blood flow is nearer 1300 ml/min than 1500 ml/min if the haematocrit is 0.40.

401.

(a) F at this level there is generalised vasoconstriction (sparing heart and brain); renal vasoconstriction leads to a marked fall in renal blood flow

(b) F it falls to zero

(c) F filtration ceases when glomerular capillary pressure is less than the sum of plasma oncotic pressure plus the hydrostatic pressure in Bowman's capsule

(d) F renal blood flow persists to much lower levels, otherwise severe renal damage would be much commoner.

402.

(a) F this would not overcome the opposing intracapsular pressure and the osmotic pressure of the plasma proteins. The value is in the range 60–70 mm Hg (8.0–9.3 kPa)

(b) F the pressure drop along the afferent arterioles (approximately 95−65 = 30 mm Hg) is less than that along the efferent arterioles to the low pressure peritubular capillaries (65−15 = 50 mm Hg)

(c) F their permeability is similar to that of most other capillaries

(d) T glomerular filtrate is a simple ultrafiltrate of the glomerular capillary blood.

403.

(a) F at low plasma levels of PAH, renal venous blood contains a negligible amount of PAH

(b) T PAH is removed from the blood in appreciable amount only by the kidneys

(c) T plasma flow $= \frac{UV}{P} = \frac{16 \times 1}{0.02} = 800$ ml/min

(d) T renal blood flow = renal plasma flow $\times \frac{1}{1\ \text{haematocrit}} =$

$800 \times \frac{1.0}{0.6} = 1333$ ml/min.

404. Renal blood flow is:

(a) reduced by about 10% when arterial pressure falls by about 10% from a value within the normal range

(b) greater per unit mass of tissue in the medulla than in the cortex

(c) reduced during emotional stress and fear

(d) determined by the metabolic needs of the kidney.

405. The cells of the distal convoluted tubule:

(a) reabsorb approximately 50% of the water in the glomerular filtrate
(b) are capable of excreting hydrogen ions by a mechanism which involves carbonic anhydrase
(c) are capable of reabsorbing sodium ions in exchange for hydrogen and potassium ions
(d) determine the final composition of urine in accordance with body needs.

406. Urea:

(a) concentration in the blood may show a ten-fold rise after a large protein meal
(b) increases the volume of urine secreted by the kidneys when its concentration in the blood rises
(c) rises in concentration as the glomerular filtrate passes down the nephron
(d) is actively secreted by the cells of the proximal and distal convoluted tubules.

407. Evacuation of the bladder:

(a) depends on the integrity of a sacral spinal reflex arc
(b) follows activation of the sympathetic nerves to the bladder

(c) is normally accompanied by reflux of bladder contents into the ureters early in micturition
(d) is prevented by destruction of the sensory nerves supplying the bladder.

404.

(a) F renal blood flow shows autoregulation; flow is relatively independent of perfusion pressure around the normal range

(b) F the reverse is true; cortical flow is several times greater than medullary flow

(c) T presumably because of circulating adrenaline and noradrenaline, together with increased activity in the vasoconstrictor nerves

(d) F the flow is vastly greater than the metabolic needs would warrant. The kidneys receive one-quarter of the cardiac output.

405.

(a) F about 80% has already been absorbed in the proximal convoluted tubule

(b) T carbonic anhydrase aids formation of carbonic acid and hence hydrogen ions

(c) T some sodium is also reabsorbed into the peritubular blood in association with chloride ions.

(d) F the collecting ducts also modify the composition of the urine.

406.

(a) F the kidney normally maintains blood urea within the range 20–40 mg/100 ml (3.3–6.7 mmol/l)

(b) T an increase in the urea load presented to the renal tubules causes an osmotic diuresis

(c) T because of the reabsorption of water that occurs from the nephron

(d) F it is reabsorbed passively by the kidney; its tubular concentration rises because relatively more water is absorbed.

407.

(a) T the reflex is under higher centre control

(b) F sympathetic fibres inhibit micturition; parasympathetic activity causes the bladder to contract

(c) F reflux does not normally occur

(d) T this breaks the reflex arc.

408. The proximal convoluted tubules:

(a) reabsorb most of the sodium in the glomerular filtrate
(b) reabsorb most of the chloride in the glomerular filtrate

(c) contain juxtaglomerular cells which secrete renin

(d) are the main target for antidiuretic hormone (ADH).

409. The figure below shows how the clearance of a substance may vary with its plasma concentration. The plasma concentration scale for line X is different from that for line Z. In this diagram the:

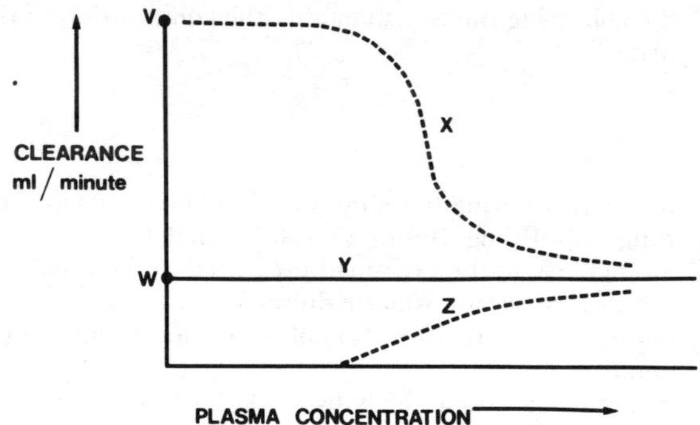

(a) line X could apply to inulin

(b) line Z could apply to glucose

(c) point V would have a value nearer 600 than 1200 ml/minute in the average adult if the renal venous concentration of substance X were zero in the initial plateau phase
(d) shape of lines X and Z, but not Y, is accounted for by active tubular transport.

408.
(a) T approximately four-fifths
(b) T the negatively charged chloride ions follow the positively charged sodium
(c) F juxtaglomerular cells are found where the afferent arteriole comes in contact with the distal convoluted tubule
(d) F ADH acts mainly on the distal parts of the nephron.

409.

(a) F line Y could apply to inulin, whose clearance is constant at W ml/minute
(b) T the clearance of glucose at low plasma levels is zero; above the renal threshold its clearance rises towards that of inulin
(c) T the clearance, at low plasma concentrations, of a substance with the characteristics of line X (PAH) corresponds to the renal *plasma* flow
(d) T in both cases, deviation from the horizontal is related to the degree by which transport maxima (secretion and reabsorption) are exceeded.

410. The collecting ducts in the kidney:

(a) can secrete water molecules actively into the urine

(b) are responsible for most of the reabsorption of water that occurs in the kidneys

(c) determine to a large extent the final osmolality of urine

(d) are rendered impermeable to water by antidiuretic hormone.

411. Aldosterone:

(a) is produced mainly in the juxtaglomerular apparatus

(b) increases sodium reabsorption by the nephron

(c) increases potassium reabsorption by the nephron

(d) tends to increase the hydrogen ion concentration in the blood.

412. The renal clearance:

(a) is calculated from UP/V where U is urinary concentration of the material, V urine volume/min, and P is plasma concentration

(b) of inulin provides an estimate of glomerular filtration rate

(c) of creatinine provides an estimate of renal plasma flow in man

(d) of phosphate is decreased by parathormone.

413. As fluid passes down the proximal convoluted tubule, the:

(a) concentrations of sulphate and uric acid fall

(b) concentrations of amino acids fall

(c) concentration of sodium falls by about four-fifths

(d) flow rate falls.

410.
(a) F no active transport systems for water have been found in the body
(b) F the proximal convoluted tubules reabsorb most of the water
(c) T by determining the amount of water leaving the duct lumen as it passes through the hypertonic medulla
(d) F the reverse is true; ADH causes an antidiuresis by allowing water to pass through the collecting duct walls into the hypertonic medullary interstitium.

411.
(a) F it is produced in the adrenal cortex
(b) T this is its primary action
(c) F the reverse; potassium is secreted in exchange for reabsorbed sodium
(d) F it tends to decrease blood hydrogen ion concentration since hydrogen ions may be secreted in exchange for the reabsorbed sodium.

412.
(a) F the formula is UV/P
(b) T no appreciable inulin is reabsorbed or secreted by the tubules, so the amount of inulin excreted in the urine is the same as the amount filtered by the glomeruli per unit time
(c) F since creatinine is treated rather like inulin in man, it is used to estimate glomerular filtration rate
(d) F it is increased; the lowered plasma phosphate level leads to a raised plasma calcium level, since the product of the calcium and phosphate ion concentrations is a constant.

413.
(a) F they rise since water reabsorption is proportionately greater than the reabsorption of sulphate or uric acid
(b) T these are reabsorbed by an active process at a proportionately greater rate than water
(c) F though about four-fifths of the sodium is reabsorbed, an equivalent amount of water is reabsorbed at the same time. Hence the sodium concentration is little changed
(d) T as water is reabsorbed the volume and therefore the rate of flow falls.

414. Potassium:

(a) is secreted in the distal convoluted tubule
(b) is reabsorbed in the proximal convoluted tubule

(c) competes with hydrogen ions for secretion by the distal convoluted tubule in exchange for sodium ions

(d) excretion in the urine is decreased by the action of mineralocorticoids.

415. Aldosterone:

(a) is secreted in increased amounts when blood volume falls

(b) is a polypeptide

(c) secretion tends to increase renal arterial pressure

(d) secretion results in a reduction in urinary volume.

416. Secretion of renin:

(a) is believed to occur from the cells of the juxtaglomerular apparatus
(b) leads to fluid retention by decreasing the glomerular filtration rate
(c) leads to increased formation of angiotensin II in the blood

(d) leads to a raised aldosterone level in systemic blood.

417. Urinary:

(a) specific gravity in normal people should not lie outside the range 1.010–1.020
(b) colour is due mainly to bile pigments
(c) pH falls as the protein content of a diet rises

(d) excretion of calcium is increased by parathormone.

414.
(a) T it is exchanged for sodium ions
(b) T most of the filtered potassium is reabsorbed in the proximal convoluted tubule so that the fluid leaving the proximal tubule is virtually potassium free
(c) T when hydrogen ions accumulate in the body, the elimination of potassium from the body by the kidneys may be reduced and result in hyperkalaemia
(d) F mineralocorticoids such as aldosterone favour sodium reabsorption and potassium secretion by the tubules.

415.
(a) T it tends to expand the extracellular fluid volume and hence blood volume by retaining sodium
(b) F as its name suggests it is a steroid, as are the other adrenal cortical hormones
(c) T the expansion in blood volume tends to increase cardiac output and hence arterial pressure
(d) T due to the increased reabsorption of sodium, chloride and water.

416.
(a) T preparations of these cells have a high renin activity
(b) F renin secretion has little effect on GFR
(c) T it converts the circulating protein angiotensinogen into angiotensin I; this is then converted into angiotensin II by a converting enzyme
(d) T via angiotensin's action on the adrenal cortex.

417.
(a) F the healthy kidney can produce big swings in urinary specific gravity e.g. 1.004–1.040
(b) F it is due to the pigment "urochrome" of uncertain origin
(c) T protein in the diet produces acid residues such as phosphate and sulphate which must be excreted by the kidney
(d) T the calcium mobilised by parathormone from the bones raises the level of calcium in the blood so that more calcium is filtered and more excreted by the kidneys.

URINARY SYSTEM
APPLIED PHYSIOLOGY

418. In chronic renal failure (equivalent to a reduction in the total number of nephrons):

(a) the specific gravity of urine is typically high (e.g. 1.030)

(b) plasma P_{CO_2} tends to be low

(c) the amount of ionised calcium in the blood rises due to calcium retention

(d) anaemia when it occurs tends to be of the iron-deficiency type.

419. A patient suffering from a deficiency of antidiuretic hormone (diabetes insipidus) is likely to:

(a) excrete urine with a specific gravity below 1.005

(b) produce a quantity of urine per day equal in volume to the glomerular filtrate

(c) have extracellular fluid with a raised osmolality

(d) have intracellular fluid with a lowered osmolality.

420. In acute tubular necrosis (swelling of renal tubule cells with loss of function and narrowing of the tubule lumen):

(a) any urine that is excreted tends to be highly concentrated

(b) the patient should not be given any fluid if he is not passing any urine

(c) the patient should not be given food with an energy content of more than about 100 calories (0.4 MJ) per day

(d) large amounts of fluid should be given intravenously at the beginning of the recovery period to increase urine production.

418.

(a) F it tends to approach 1.010, varying less as the disease progresses

(b) T the metabolic acidosis in chronic renal failure causes hyperventilation

(c) F the level of ionised calcium tends to fall because of phosphate retention (the |calcium | |phosphate | product remaining constant) and failure of the kidney to hydroxylate, and so activate, vitamin D

(d) F it is normocytic, normochronic, indicating inadequate marrow activity probably due to erythropoietin deficiency.

419.

(a) T it is typically in the range 1.000–1.005

(b) F the volume is much less, 5–20 litres per day; at least 80% of the glomerular filtrate (about 180 litres per day) is reabsorbed in the absence of ADH

(c) T due to relatively greater loss of water than of salt by the kidneys

(d) F the raised osmolality of the ECF draws water from the intracellular compartment; death if it occurs results from intracellular dehydration.

420.

(a) F what little glomerular filtrate can pass through the tubules is not modified, its specific gravity equals that of plasma

(b) F water lost via the lungs and skin (about 1 litre) needs to be replaced

(c) F the diet should consist of enough carbohydrate to reduce body protein breakdown to a minimum, e.g. 200 g lactose per day (800 calories; 3.2 MJ)

(d) F such treatment might cause fatal pulmonary oedema.

421. The diagram below represents the relationship between bladder volume and pressure for a number of bladder conditions. If curve Y represents the normal bladder in the relaxed state, and curve V represents the normal bladder in the contracted state, then:

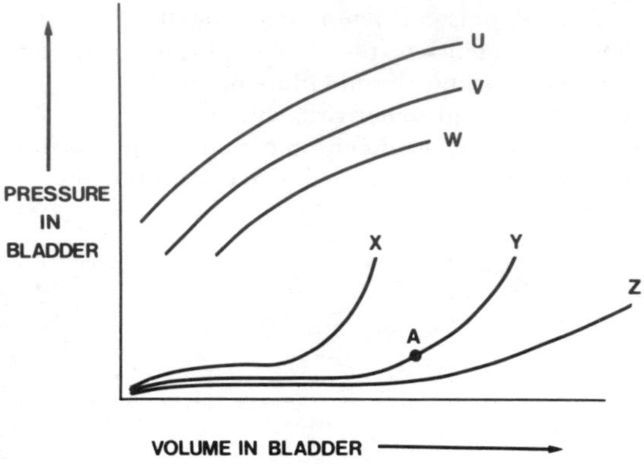

(a) point A would correspond to a bladder volume closer to 100 than 500 ml
(b) curve U rather than curve W would represent the bladder of a patient whose urethra had been partially obstructed for several months
(c) curve X rather than curve Z would represent the bladder when inhibition of the spinal micturition reflex has been removed
(d) curve Z would represent the bladder with chronic infection.

422. When a patient is treated with an aldosterone antagonist there is likely to be a fall in:
(a) urine volume
(b) plasma potassium concentration
(c) blood volume
(d) blood viscosity.

421.

(a) F the normal relaxed bladder accommodates about 500 ml before the pressure begins to rise rapidly
(b) T due to compensatory hypertrophy the bladder tends to exert a greater than normal pressure at a given degree of stretch
(c) T the uninhibited neurogenic bladder shows a greater than normal pressure rise as its volume increases
(d) F the increased sensitivity of the sensory nerves in the bladder wall to stretch when the bladder is infected causes reflex emptying at low volumes.

422.

(a) F it increases since sodium reabsorption is reduced
(b) F the potassium level rises (aldosterone increases the secretion of potassium by the tubules into the urine)
(c) T due to a fall in extracellular sodium and hence water
(d) F the viscosity increases as the haematocrit increases.

423. The rate of urine flow when the bladder is being emptied:

(a) is normally higher in men than in women

(b) is normally nearer 5 ml than 20 ml per second

(c) tends to rise when the prostrate gland is enlarged
(d) would tend to fall after administration of an anticholinergic drug.

424. In the treatment of chronic renal failure by dialysis, the dialysis fluid should:

(a) contain urea at a concentration around half that in the patient's plasma
(b) usually have the same sodium:potassium ratio as normal plasma

(c) be hypertonic to normal extracellular fluid if the dialysis is to lower the patient's blood pressure
(d) normally correct the patient's anaemia (if present) provided its electrolyte composition is appropriate.

425. Renal transplantation:

(a) needs to be accompanied by immunosuppression even if the donor is the patient's identical twin

(b) can be judged successful if the glomerular filtration rate is 10 ml per minute in an adult
(c) can be expected to restore abnormal calcium and phosphorous metabolism towards normal in most cases
(d) can be expected to cure anaemia in most cases where there is no deficiency of raw materials nor disease of bone marrow.

426. Long-standing obstruction of the urethra, e.g. by an enlarged prostrate gland, may result in:

(a) hypertrophy of the bladder muscle

(b) dilatation of the ureters

(c) a reduced glomerular filtration rate and eventual renal failure
(d) infection of the urinary tract.

423.

(a) F it is higher in women, since their outflow resistance is lower

(b) F it normally exceeds 20 ml per second in both sexes (e.g. 300 ml in less than 15 seconds)

(c) F it falls due to the increased outflow resistance

(d) T due to reduction in the strength of contraction of the bladder.

424.

(a) F the dialysate is urea-free to allow the maximal concentration gradient for transfer from the patient

(b) F the dialysate usually contains little or no potassium, since potassium excess is a common problem in renal failure

(c) T this tends to draw fluid out of the patient's plasma and hence lower his blood volume and blood pressure

(d) F anaemia in renal failure cannot usually be corrected by dialysis; at least part of the explanation seems to be lack of erythropoietin.

425.

(a) F the case of the identical twin donor is an exception to the general rule, since donor and recipient share the same genetic make-up

(b) F the value should be much nearer the normal of 120–150 ml per minute

(c) T this tends to lead to gradual healing of areas of bone damage

(d) T the anaemia resolves in 6 to 12 weeks; in short, transplantation usually reverses all the abnormalities of renal failure.

426.

(a) T due to the increased work it has to do in overcoming the obstruction

(b) T if the rise in bladder pressure is sufficient to overcome the ureterovesical valves

(c) T due to the back pressure in the tubules

(d) T an inability to evacuate the bladder completely allows urine to stagnate in the bladder and permits the multiplication of micro-organisms.

427. Samples taken from the ureters of a patient with severe narrowing of the renal artery on one side gave the results shown below. Plasma creatinine level was 1 mg/100 ml and PAH level was 3 mg/100 ml. In this patient:

	left ureter	right ureter
urine volume (ml/min)	0.2	6.0
creatinine conc. (mg/100 ml)	100	10
PAH conc. (mg/100 ml)	1000	150

(a) glomerular filtration rate (creatinine clearance) was 10 times as great on the left side as on the right
(b) renal plasma flow was approximately 67 ml/min on the left
(c) renal blood flow was 900 ml/min on the right (if the haematocrit was 33 %)
(d) the right kidney had the narrowed renal artery.

428. Interference with active reabsorption of sodium by the kidney is likely to cause:
(a) a rise in the volume of urine passed in a given time
(b) a rise in plasma potassium level
(c) a decreased interstitial fluid volume
(d) a rise in plasma specific gravity.

429. A patient treated with a drug which inhibits carbonic anhydrase is likely to have:
(a) a fall in plasma bicarbonate
(b) decreased loss of hydrogen ions in the urine
(c) decreased loss of potassium in the urine
(d) an increased urinary volume.

430. A patient suffering from chronic renal failure typically:
(a) exacerbates his pH abnormality by vomiting
(b) exacerbates his pH abnormality if he increases his intake of meat
(c) has a low rate of clearance of creatinine and inulin
(d) has a raised blood level of uric acid.

427.

(a) F creatinine clearance (UV/P) was 20 ml/min on the left and 60 ml/min (normal) on the right)
(b) T from PAH clearance—a very low value
(c) F blood flow = plasma flow × 1/1 − Ht = 300 × 3/2 = 450 ml/min
(d) F the left side had the abnormality.

428.

(a) T extra water is lost with the extra sodium and chloride lost
(b) F the passive reabsorption of K^+ depends on the active reabsorption of Na^+; the plasma level of K^+ may fall dangerously
(c) T hence oedema is relieved
(d) T due to concentration of the proteins by removal of water.

429.

(a) T bicarbonate formation in the tubules is decreased
(b) T hydrogen ion formation in the tubules is also decreased
(c) F potassium loss is increased since potassium rather than hydrogen ions are excreted by the distal tubules in exchange for sodium
(d) T the filtered sodium bicarbonate is not reabsorbed and causes an osmotic diuresis.

430.
(a) F vomiting gastric acid would tend to improve the typical acidosis
(b) T protein metabolism leads to acid end-products which are incompletely excreted in renal failure
(c) T these indicate the reduction in glomerular filtration rate
(d) T as with the other end-products of protein digestion.

431. Cutting the sympathetic nerves to the bladder may cause:
(a) retention of urine
(b) loss of pain sensation in the bladder
(c) infertility in the male
(d) dilatation of the external sphincter of the bladder.

432. Sudden (acute) renal failure differs from gradual (chronic) renal failure in that:
(a) potassium retention tends to be more severe
(b) depression of the bone marrow is not seen
(c) it does not cause a metabolic acidosis
(d) there is less need for protein restriction in the diet.

433. In the treatment of a patient with chronic renal failure:
(a) protein should be excluded from the diet
(b) water intake should be restricted to about 0.5 litre per day
(c) the diet should contain iron supplements to prevent anaemia
(d) the diet should be potassium free.

431.
(a) F in some people frequency of micturition is increased
(b) T the afferent fibres carrying pain sensation from the bladder run with sympathetic nerves
(c) T sympathetic fibres are necessary for the closure of the internal sphincter of the bladder during ejaculation thereby preventing reflux of seminal fluid into the bladder
(d) F the external sphincter is composed of skeletal muscle and is innervated by the pudendal (somatic) nerves.

432.
(a) T potassium retention is one of the greatest hazards in acute renal failure and may result in death from myocardial depression
(b) F bone marrow depression may be severe in acute renal failure resulting in anaemia, susceptibility to infection and a tendency to bleed
(c) F acute renal failure results in the rapid development of a severe metabolic acidosis with hyperventilation (Kussmaul breathing) since the kidneys lose the ability to manufacture bicarbonate
(d) F protein restriction is advisable in both cases.

433.
(a) F this would result in increased utilisation of the patient's own proteins; protein restriction is, of course, necessary
(b) F water intake may need to be increased to about 3 litres per day; since the kidneys lose the ability to concentrate urine, larger urinary volumes are needed to eliminate waste products of metabolism
(c) F the anaemia in chronic renal failure is due to bone marrow depression and does not respond to iron
(d) F except in terminal failure, potassium retention is not a problem; a potassium-free diet might result in potassium depletion.

434. A long-standing increase in the P_{CO_2} of arterial blood is associated with:

(a) an increase in plasma bicarbonate

(b) a decrease in the amount of ammonium salts in the urine

(c) a decrease in the plasma concentration of potassium ions

(d) an increase in the ratio of monohydrogen to dihydrogen phosphate in the urine.

434.

(a) T this is the normal renal compensation for a respiratory acidosis

(b) F the reverse is true; in acidosis, tubular cells excrete more NH_3 to buffer the excreted hydrogen ions

(c) F since hydrogen ions compete with potassium ions for excretion, acidosis tends to cause potassium retention

(d) F there is an increase in the ratio of dihydrogen to monohydrogen phosphate as hydrogen ions are taken up by the phosphate buffer system.

ENDOCRINE SYSTEM
BASIC PHYSIOLOGY

435. In the plasma, the half-life of:

(a) a substance is half the time it takes for the concentration to fall from its initial level to an undetectable level

(b) insulin is between five and ten hours

(c) thyroxine is greater than that of insulin

(d) thyroxine (T4) is less than that of triiodothyronine (T3).

436. During sleep, as compared with the waking state, there is usually a higher circulating level of:

(a) cortisol

(b) insulin

(c) antidiuretic hormone

(d) adrenaline.

437. Adrenocorticotrophic hormone (ACTH):

(a) is produced by neurones whose cell bodies lie in the hypothalamus

(b) has a greater effect on aldosterone secretion than on cortisol secretion

(c) output tends to increase when the circulating cortisol level falls

(d) production is controlled by a releasing hormone from the hypothalamus.

438. Thyroid hormones tend to increase:

(a) peripheral resistance

(b) the frequency of defaecation

(c) the energy expenditure required to perform a given amount of work

(d) the duration of tendon reflexes.

435.
(a) F it is the time it takes for the concentration to fall to half the original level.
(b) F it is much shorter (about five minutes); if its action were not terminated abruptly when blood glucose declined, there would be serious hypoglycaemia
(c) T its action is much more prolonged, due largely to its very high degree of protein binding (about 99·9%)
(d) F T4 is more highly protein bound and has the longer half-life.

436.
(a) F this catabolic hormone is secreted predominantly during the waking hours
(b) F insulin secretion occurs mainly in association with meals
(c) T gradual dehydration occurs during sleep; the decreased urine production reduces the need to empty the bladder during the night
(d) F secretion of adrenaline is associated with the stresses of waking activity.

437.
(a) F it is produced by gland cells in the anterior pituitary
(b) F aldosterone production is largely controlled by angiotensin
(c) T this tends to maintain the blood level of cortisol by a negative feedback system
(d) T corticotrophin releasing hormone (CRH) is part of the feedback control loop.

438.
(a) F this falls due to peripheral vasodilatation generated by the increase in metabolism
(b) T excessive thyroid activity may cause diarrhoea; the mechanism is obscure
(c) T thyroid hormones tend to uncouple oxidation from phosphorylation in the mitochondria so that more oxidation is needed to produce sufficient utilisable energy
(d) F the reverse is true.

439. Aldosterone production is increased by a fall in:

(a) plasma osmolality

(b) blood volume
(c) the plasma renin concentration
(d) renal blood flow.

440. Glucocorticoids in the blood:

(a) inhibit the pituitary production of ACTH

(b) promote sodium reabsorption by the kidney
(c) decrease the total white cell count

(d) stimulate growth in lymphoid tissue.

441. The diagram shows trends which may follow when a normal person receives slow intravenous infusions of noradrenaline (Nor) and adrenaline (Adr). The trends indicated could represent changes in:

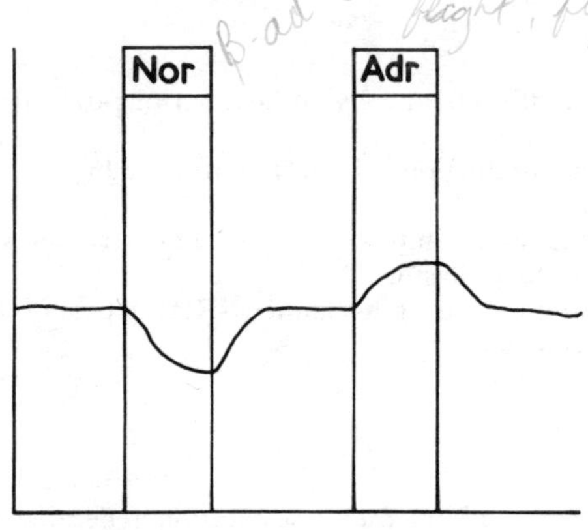

(a) systolic blood pressure
(b) diastolic blood pressure
(c) total peripheral resistance
(d) forearm blood flow.

439.
(a) F this leads to decreased production of ADH by the posterior pituitary
(b) T sodium is retained and the blood volume restored
(c) F it is increased when the plasma renin increases
(d) T this results in renin secretion by the juxtaglomerular apparatus with the subsequent formation of angiotensin which stimulates aldosterone formation.

440.
(a) T this is the negative feedback system regulating the blood glucocorticoid level
(b) T glucocorticoids have some mineralocorticoid action
(c) F they raise the WCC, mainly by increasing the neutrophil count; the number of eosinophils falls
(d) F glucocorticoids can be used to suppress immune responses.

441.

(a) F systolic pressure is raised by both catecholamines
(b) F diastolic pressure is raised by noradrenaline
(c) F noradrenaline increases total peripheral resistance
(d) T adrenaline causes vasodilatation whereas noradrenaline causes vasoconstriction in skeletal muscle.

442. The secretion of growth hormone:
(a) is normally dependent on the integrity of the hypothalamus

(b) ceases when the adult state is reached

(c) is decreased following severe injury (trauma)
(d) is increased during fasting and other states where the blood glucose level is reduced.

443. Parathormone:
(a) is produced by the P cells of the thyroid gland
(b) is a steroid
(c) raises the serum phosphate level

(d) decreases calcium excretion by the kidney.

444. Antidiuretic hormone (vasopressin):
(a) reduces the rate of glomerular filtration
(b) is secreted at a rate determined by plasma osmolality
(c) increases the permeability of renal collecting duct cells to water

(d) is secreted by nerve cells with their cell bodies in the hypothalamus.

445. Pancreatic glucagon:
(a) is a polypeptide produced by the beta cells of the islets of Langerhans
(b) is essential for the maintenance of the blood glucose level

(c) output is inversely proportional to the blood glucose level

(d) increases the breakdown of liver glycogen.

442.
(a) T the hypothalamus produces a growth hormone releasing hormone which is carried to the pituitary by the pituitary portal system
(b) F the blood levels of growth hormone are similar in children and adults
(c) F stress results in increased secretion
(d) T its action in raising blood glucose might be useful here.

443.
(a) F it is produced in the parathyroid glands
(b) F it is a polypeptide (MW around 8000)
(c) F it lowers it by reducing phosphate reabsorption; calcium is then mobilised from bone to maintain the solubility product |calcium |. |phosphate | constant
(d) F as the plasma calcium level rises in hyperparathyroidism, more calcium is excreted in the urine; and bones may be depleted of calcium.

444.
(a) F it has no effect on the GFR
(b) T osmolality is detected by hypothalamic osmoreceptors
(c) T hence increasing water reabsorption by allowing water to pass into the hypertonic fluid of the medullary interstitium
(d) T the axons of these cells pass into the posterior pituitary and release their secretions there.

445.
(a) F it is produced by the alpha cells; insulin is produced by the beta cells
(b) F the blood glucose level can be maintained without glucagon; e.g. in patients who have had their pancreas removed, insulin replacement only is needed
(c) T as with insulin secretion there seems to be a direct feedback control by the blood glucose level
(d) T as does adrenaline.

446. The serum calcium:

(a) concentration of about 10 mg/100 ml may be expressed as 2.5 mmol/l (atomic weight of calcium = 40)

(b) concentration influences the rate of parathormone production by an action on the hypothalamus

(c) is approximately 50±10% ionised

(d) becomes less ionised when blood pH falls.

447. Cortisol:

(a) secretion increases following an injury to the body such as the breaking of a leg

(b) favours protein synthesis (anabolism)

(c) enhances the effects of antibody-antigen reactions

(d) tends to lower the blood pressure.

448. When a thyroid gland increases its activity:

(a) the follicles enlarge and fill with colloid

(b) small gaps (lacunae) appear in the colloid

(c) the follicular cells become more columnar

(d) it takes up iodide from the blood at a faster rate.

449. The hypothalamus produces releasing hormones which influence the release of:

(a) thyroxine from the thyroid gland

(b) oxytocin from the posterior pituitary gland

(c) glucagon from the pancreas

(d) testosterone from the testis.

446.
(a) T since the atomic weight of calcium is 40, one mmol is 40 mg. One litre of serum contains 100 mg which is equivalent to 2.5 mmol
(b) F the serum calcium regulates the output of parathormone by a direct action on the parathyroid glands
(c) T most of the rest is bound to protein, mainly albumin
(d) F a fall in pH reduces the ability of protein to bind calcium, thereby raising the free calcium level.

447.
(a) T the chances of survival after trauma are reduced in adrenal failure
(b) F it favours protein breakdown (catabolism)
(c) F it suppresses these effects
(d) F it tends to raise blood pressure; one of the mechanisms involved in the expansion of the blood volume that results from increased sodium reabsorption by the kidneys.

448.
(a) F colloid is emptied from the follicles during activity as the stored thyroxine is liberated
(b) T this suggests breakdown of thyroglobulin and release of hormone
(c) T a flattened epithelium suggests a less active state
(d) T the rate of uptake of radioactive iodide is an index of thyroid activity.

449.
(a) T thyrotropin-releasing hormone (TRH)
(b) F oxytocin is released at the terminals of nerves whose cell bodies lie in the hypothalamus
(c) F this is controlled by the blood glucose level
(d) T luteinising hormone/follicle stimulating hormone-releasing hormone (LH/FSH-RH) regulates luteinising hormone (LH) release and hence testosterone secretion.

450. Release of the catecholamines adrenaline and noradrenaline from the adrenal medulla increases:

(a) the blood level of glucose by favouring glycogenolysis in liver and muscle cells

(b) the blood levels of free fatty acids and ketone bodies

(c) muscle and splanchnic blood flow

(d) when the parasympathetic nerves to the adrenal gland are cut.

451. Thyroid-stimulating hormone (TSH) is secreted at a higher rate:

(a) after removal of one lobe of the thyroid gland

(b) in warm-adapted than in cold-adapted subjects

(c) when the basal metabolic rate falls

(d) when the diet is deficient in iodine.

452. Insulin:

(a) stimulates the release of free fatty acid from adipose tissue

(b) secretion tends to cause the blood potassium level to rise

(c) facilitates entry of glucose, amino acids and fatty acids into skeletal muscle cells

(d) secretion increases when the vagus nerves are stimulated.

453. The pituitary gland:

(a) consists of an anterior lobe developed from the alimentary tract and a posterior lobe developed from the brain

(b) has a posterior lobe which is largely influenced by hormones passing down blood vessels in the pituitary stalk

(c) has a nervous connection with the brain

(d) exerts its effects only by influencing other endocrine glands.

450.

(a) T adrenaline does this by activating phosphorylase

(b) T the mobilisation of these substances from the liver and fat depots provides substrates that are readily available to provide energy

(c) F muscle vessels dilate because of their predominant beta receptors; splanchnic vessels constrict because of their predominant alpha receptors

(d) F the adrenal medulla does not have a parasympathetic supply; catecholamine release is controlled by "preganglionic" sympathetic fibres.

451.

(a) T due to a reduction in circulating thyroxine

(b) F heat-adapted people have decreased thyroid activity

(c) F TSH and thyroxine influence the basal metabolic rate, not *vice versa*

(d) T the inability of the thyroid to manufacture sufficient thyroxine in iodine deficiency results in increased THS secretion.

452.

(a) F it inhibits free fatty acid release and so reduces ketone body formation in the liver

(b) F the blood level of potassium tends to fall; insulin promotes K^+ entry into muscle cells

(c) T in general, it increases the permeability of cell membranes to small organic molecules

(d) T this may increase insulin secretion when food is eaten and limit the resulting rise in blood glucose.

453.

(a) T this correlates with the structure of the adult gland

(b) F this refers to the anterior lobe and the pituitary portal system

(c) T the nerve fibres of the stalk provide such a connection

(d) F growth hormone and antidiuretic hormone, for example, influence tissues directly.

454. Thyrocalcitonin:
(a) lowers the basal metabolic rate
(b) reduces the blood calcium level in the absence of the parathyroid glands
(c) is released when the blood phosphate level rises
(d) is produced in the thyroid gland.

455. The thyroid gland:
(a) takes up iodide against its electrochemical gradient
(b) is regulated by a hormone released from the posterior pituitary gland
(c) is relatively avascular and bleeds little when cut
(d) contains enzymes which oxidise iodide to iodine.

456. Adrenaline differs from noradrenaline in that it:
(a) lacks a methyl (CH_3) group on its nitrogen atom
(b) causes tachycardia when injected intravenously
(c) is the catecholamine secreted in greatest amount by the adrenal medulla
(d) increases the strength of myocardial contraction.

457. Vitamin D:
(a) increases the intestinal absorption of calcium
(b) is essential for normal calcification of the proliferating cartilage at epiphyses
(c) is inactive before it is hydroxylated in the kidney
(d) may be formed in the body.

458. Prolactin:
(a) is synonymous with luteinising hormone
(b) is the hormone mainly responsible for the growth of breast tissue
(c) secretion reaches a peak around the time of delivery (parturition)
(d) release is inhibited by dopamine.

454.
(a) F it does not affect the basal metabolic rate
(b) T this indicates that it does not act through the parathyroids
(c) F it is released when the blood calcium level rises
(d) T by parafollicular cells derived embryologically from the ultimobranchial bodies which become incorporated in the thyroid.

455.
(a) T it is taken up by an active process (the iodide pump); the ratio of iodide in thyroid and plasma is about 25:1
(b) F it is regulated by thyroid-stimulating hormone (TSH) released by the anterior pituitary
(c) F the thyroid has one of the highest rates of flow per gram of tissue of any organ (about 50 times that of resting skeletal muscle)
(d) T this is a stage in the formation of thyroxine which is synthesised on thyroglobulin molecules in the follicle.

456.
(a) F it is noradrenaline which lacks the methyl group (NOR = nitrogen ohne radicale)
(b) T noradrenaline injections cause a reflex bradycardia
(c) T 80% of catecholamine secretion by the human medulla is adrenaline
(d) F both increase the strength of myocardial contraction.

457.
(a) T the mechanism is not known
(b) T in its absence developing bones are weak and deformed (rickets)
(c) T the 1,25 hydroxylated form is released by the kidney in accordance with the body need for vitamin D
(d) T by the action of sunlight on ergosterol in the skin.

458.
(a) F luteinising hormone (LH) is one of the pituitary gonadotrophins
(b) F breast development is mainly due to oestrogen and progestogen secretion
(c) T it is thought to be responsible for the initiation of lactation
(d) T hypothalamic release of dopamine may be a mechanism regulating prolactin secretion.

459. The amount of ionised calcium in blood falls:

(a) if sodium citrate is added to it
(b) if the thyroid gland is perfused with a calcium-rich solution
(c) in hyperventilation
(d) when the blood phosphate level falls.

460. Thyroxine:

(a) is synthesised from two tyrosine units and three atoms of iodine
(b) accounts for nearer 60% than 90% of circulating thyroid hormones
(c) circulates in the blood mainly as free thyroxine
(d) increases the resting rate of carbon dioxide production.

461. Parathormone:

(a) decreases the renal clearance of phosphate
(b) mobilises calcium from bone independently of its action on the kidney
(c) depresses the activity of the anterior pituitary gland
(d) in the blood is increased in amount when the calcium level falls.

462. The chemical structure of insulin:

(a) is identical in all mammalian species
(b) is such that it is effective when taken by mouth
(c) consists of three chains of amino acids linked by disulphide bridges
(d) has been synthesised in the laboratory.

463. The chief hormones secreted by the adrenal cortex:

(a) include cortisol, corticosterone, cholesterol, aldosterone and the androgen, dehydroepiandrosterone
(b) are largely bound to plasma proteins
(c) are excreted mainly in the bile after conjugation in the liver
(d) may be divided functionally into mineralocorticoid, glucocorticoid and sex corticoid groups.

459.
(a) T calcium ions are bound as calcium citrate
(b) T due to the release of thyrocalcitonin
(c) T the resulting respiratory alkalosis increases the calcium binding power of the plasma proteins
(d) F the product of the concentrations of calcium and phosphate ions is constant (solubility product).

460.
(a) F this is triiodothyronine; thyroxine is formed from two tyrosine units and four iodine atoms
(b) F over 90%
(c) F about 99·9% is bound to plasma proteins
(d) T by increasing the basal metabolic rate.

461.
(a) F renal clearance is increased, probably because of a fall in the Tm for phosphate reabsorption
(b) T it does so in bone tissue culture and in animals whose kidneys have been removed
(c) F the anterior pituitary is not affected by parathyroid function
(d) T blood calcium level determines the rate of secretion of parathormone.

462.
(a) F minor differences occur but mammalian insulin can be used for replacement therapy in man
(b) F as a polypeptide it is broken down by the digestive juices
(c) F it has only two chains
(d) T in 1964 by Katsoyannis.

463.
(a) F cholesterol is not secreted and is not a hormone; the others are correct
(b) T for example, the alpha globulin, transcortin, binds cortisol in the plasma
(c) F they are conjugated in the liver but are then excreted by the kidney, e.g. urinary 17-ketosteroids
(d) T the functions of the first two groups overlap.

464. The islets of Langerhans:

(a) contain beta cells whose granules fix anti-insulin antibodies
(b) contain alpha cells which secrete glucagon
(c) have a high zinc content relative to other body tissues
(d) contain agranular gamma cells which are thought to secrete a hormone which controls the blood glucose level.

465. Thyroxine:

(a) formation requires the amino acid alanine
(b) is essential for the development of the brain
(c) is essential for adequate red cell production
(d) has an effect which is slower in onset and longer lasting than that of triiodothyronine (T3).

466. The adrenal cortex:

(a) produces the hormone aldosterone when the sympathetic nerves to the gland are stimulated
(b) is a source of sex hormones
(c) is influenced by an anterior pituitary hormone
(d) is the main body source of angiotensin.

467. Growth hormone:

(a) contains amino groups
(b) levels in the blood are raised when the blood glucose levels are raised, e.g. by ingestion of a meal
(c) has a lactogenic effect
(d) increases the size of the viscera.

464.

(a) T this is evidence that these cells secrete insulin

(b) T alpha and beta cells can be distinguished using special stains

(c) T insulin has an affinity for zinc and may be stored in zinc aggregates

(d) F the function of these cells is not known.

465.

(a) F it is formed from 2 tyrosine molecules and 4 iodine atoms

(b) T lack of thyroxine in infancy causes cretinism

(c) T deficiency causes anaemia

(d) T onset in 2 – 3 days compared with 12 – 24 hours; maximal effect in 10 days compared with 2 – 3 days.

466.

(a) F the sympathetic nerves to the adrenals supply the medullary cells

(b) T these appear to control the growth of axillary and pubic hair

(c) T adrenocorticotrophic hormone (ACTH) regulates mainly glucocorticoid production

(d) F angiotensin is formed in the blood stream by the action of renin and acts on the adrenal cortex to release aldosterone.

467.

(a) T it is a polypeptide

(b) F the blood level of growth hormone falls when the blood glucose level rises

(c) T growth hormone and prolactin are proteins of similar structure produced by acidophil cells

(d) T growth hormone seems to be able to stimulate growth of nearly all tissues.

468. The level of thyroid-stimulating hormone (TSH):
(a) in the pituitary portal blood is the main factor controlling thyroxine production
(b) leaving the pituitary is reduced by an increase in the level of thyroxine in the general circulation
(c) reaching the thyroid gland increases when the individual moves to a colder environment
(d) in the blood is influenced by a releasing factor from the hypothalamus.

469. Adrenal medullary hormones:
(a) include isoprenaline
(b) stimulate both alpha and beta adrenoceptors
(c) can be inactivated by monoamine oxidase
(d) are essential for life.

470. After the ingestion of 50 g glucose by a normal subject who has been on a 300 g carbohydrate diet for 3 days, the blood level of:
(a) glucose should rise by not more than 10%
(b) glucose reaches, on average, the same maximum (±10%) as when the same amount of glucose is infused intravenously, e.g. over 30 minutes
(c) insulin reaches a higher maximum than when the same amount of glucose is given intravenously
(d) glucose after the initial peak may fall to a trough below the pre-test level.

471. Secretion of growth hormone:
(a) decreases during sleep
(b) decreases during starvation
(c) tends to result in a positive potassium balance in children
(d) tends to result in a positive sodium balance in children.

468.
(a) F it is the level in the thyroid arteries which controls thyroxine production
(b) T this is the feedback mechanism which regulates basal thyroxine level via the anterior pituitary
(c) T this raises the circulating thyroxine level and increases heat production
(d) T thyrotropin-releasing factor (TRF) is carried from the hypothalamus to the pituitary by the pituitary portal vessels.

469.
(a) F this catecholamine is not found in the body; it is made synthetically
(b) T however, adrenaline has the greater affinity for beta receptors
(c) T also by catechol-o-methyl transferase
(d) F their functions can be duplicated by sympathetic nerves.

470.
(a) F it may rise by 50% or more, e.g. 80 → 120 mg/100 ml (4.5 → 6.75 mmol/l)
(b) F intravenous injection leads to a much higher level, e.g. twice as high
(c) T the level is much higher after oral glucose, suggesting that the digestive process, with increased levels of gut hormone and of vagal activity, augments insulin release
(d) T this is due to the increased insulin secretion provoked by the high blood glucose level.

471.
(a) F it increases, and may be responsible for the anabolic repair of the catabolic activities of the day
(b) F it increases, probably in response to the fall in mean blood glucose level
(c) T potassium is retained to meet the intracellular cation needs of the induced tissue growth
(d) T sodium is retained as the volume of extracellular fluid increases.

472. Adrenaline differs from noradrenaline in that it:
(a) is a more potent dilator of the bronchi

(b) reduces peripheral resistance

(c) constricts blood vessels in mucous membranes

(d) increases the blood glucose level to a greater extent.

473. The effects of oxytocin:
(a) on the uterus are potentiated by oestrogen
(b) on the uterus are potentiated by progesterone
(c) in high doses include antidiuresis

(d) on the breast include direct stimulation of milk formation.

474. Secretin differs from cholecystokinin-pancreozymin (CCK-PZ) in that it:
(a) is secreted into the blood by mucosal cells in the small intestine

(b) stimulates the pancreas to produce a juice rich in enzymes

(c) has less effect on smooth muscle in the wall of the gall bladder

(d) decreases gastric motility.

472.

(a) T it has a stronger beta stimulant effect than noradrenaline (bronchodilatation is a beta effect)

(b) T it increases cardiac output without having much effect on mean arterial pressure; noradrenaline produces a substantial rise in mean arterial pressure

(c) F both constrict blood vessels in mucous membranes and have a decongestive effect

(d) T adrenaline with its greater beta effect is more potent in this respect.

473.

(a) T oestrogen makes uterine smooth muscle more excitable

(b) F progesterone makes uterine smooth muscle less excitable

(c) T the effects of the two posterior pituitary hormones tend to overlap

(d) F oxytocin causes milk to be expressed from the lactating breast by contracting the myoepithelial cells which surround the breast alveoli.

474.

(a) F both are secreted by these cells

(b) F CCK-PZ results in the production of such a juice; secretin juice is poor in enzymes but rich in bicarbonate

(c) T both hormones stimulate the gall bladder muscle but CCK-PZ does so more strongly

(d) T CCK-PZ increases gastric motility whereas secretin depresses it.

ENDOCRINE SYSTEM
APPLIED PHYSIOLOGY

475. Treatment with a drug which inhibits the conversion of angiotensin I to angiotensin II tends to increase rather than decrease:

(a) systemic arterial blood pressure

(b) plasma renin level
(c) the circulating level of angiotensin I

(d) total body potassium.

476. The diagram below shows some measurements of plasma adrenocorticotrophic hormone (ACTH) levels in three patients, X, Y, and Z. In this diagram the:

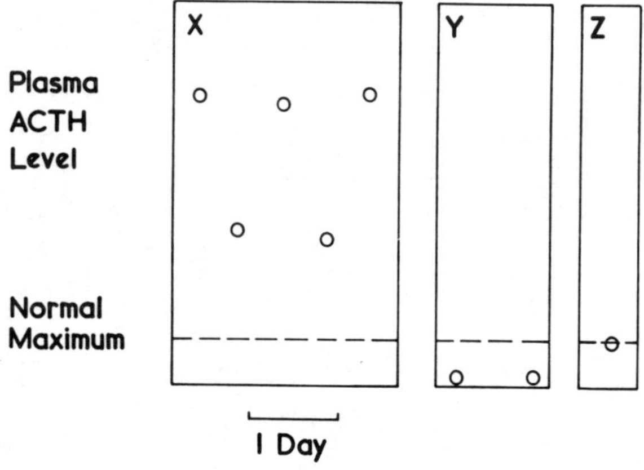

(a) results for patient X show a rhythm of normal frequency/cycle length but of abnormally great amplitude of variation
(b) results for patient X are consistent with failure of cortisol secretion due to a pituitary abnormality

(c) results for patient Y are consistent with suppression of secretion of ACTH and cortisol due to administration of a high level of exogenous glucocorticoid
(d) result for patient Z would be more suggestive of primary failure of cortisol secretion if the blood sample had been taken at 3 a.m. rather than 9 a.m.

475.

(a) F blood pressure falls (angiotensin II raises blood pressure by a variety of mechanisms, e.g. constriction of resistance vessels and increase in blood volume through release of aldosterone)

(b) T the fall in blood pressure increases release of renin

(c) T release of renin increases formation of angiotensin I, whose conversion to angiotensin II is then blocked

(d) T the fall in the level of angiotensin II leads to reduced aldosterone secretion and hence reduced potassium excretion.

476.

(a) T cycle length is about one day (circadian) but amplitude is about three times the normal maximal value

(b) F the very high levels of ACTH suggest that the failure lies primarily in the adrenals, the high ACTH levels being due to removal of feedback inhibition due to poor output of cortisol by the adrenals

(c) T in such a case the enormous negative feedback suppresses secretion of ACTH (and hence of cortisol)

(d) T such a level at 3 a.m. would be definitely raised, whereas at 9 a.m. (the time of the normal maximum) the result is equivocal.

477. A patient with hypothyroidism is likely to have:

(a) a subnormal mouth temperature
(b) a tendency to fall asleep frequently
(c) increased body hair (hirsutism)

(d) moist hands and feet.

478. A patient who develops sudden parathyroid deficiency is likely to:

(a) get skeletal muscle spasms
(b) die if no treatment is given
(c) be improved by a slow intravenous injection of calcium ions, e.g. calcium gluconate
(d) be improved in the long-term by regular doses of vitamin D.

479. The graph shows some urinary measurements in a patient with diabetes insipidus who was treated once daily with a long-acting preparation of antidiuretic hormone (ADH). In this case:

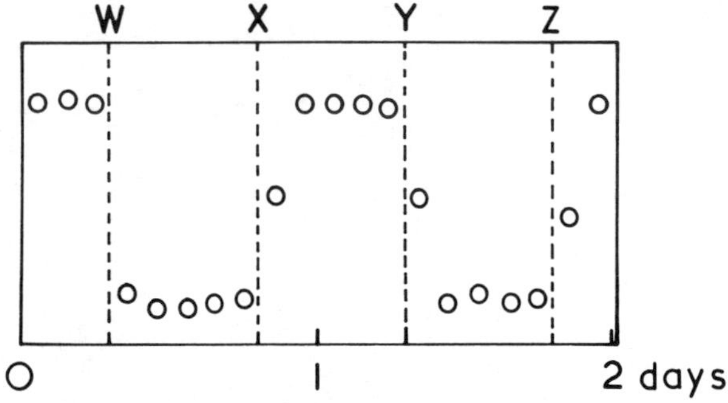

(a) if the results indicate urinary volume, then the ADH was administered at points W and Y, rather than at points X and Z
(b) if the results indicate urinary osmolality, then the ADH was administered at points W and Y, rather than at points X and Z

(c) urinary flow prior to administration of ADH would be nearer 5 than 1 ml per minute

(d) an effective dose of ADH could be expected to change urinary osmolality from less than 50% that of plasma to more than 150% that of plasma.

477.

(a) T due to a low metabolic rate

(b) T due to impaired brain function

(c) F generalised hair loss is a characteristic early sign of hypothyroidism

(d) F the skin is dry, coarse and cold; sweating is below average in hypothyroidism.

478.

(a) T this is tetany

(b) T from convulsions due to severe tetany

(c) T this is the treatment of choice

(d) T this helps to raise plasma calcium by increasing intestinal absorption of calcium.

479.

(a) T on each occasion the urinary volume fell dramatically after administration of ADH.

(b) F ADH causes greatly increased reabsorption of water rather than salt in the collecting ducts, so that urinary volume falls and osmolality rises

(c) T 1 ml per minute is normal (equivalent to 1440 ml per day); in diabetes insipidus the rate is typically around 5 ml per minute (7200 ml per day)

(d) T e.g. from 100 to 600 mOsmol/kg H_2O (plasma osmolality is normally slightly under 300 mOsmol/kg H_2O).

480. A patient suffering from severe uncontrolled diabetes mellitus is likely to have a raised:

(a) blood potassium level

(b) urine specific gravity e.g. 1·040
(c) blood volume
(d) arterial P_{CO_2}.

481. Hyperthyroidism is associated with:

(a) a positive nitrogen balance

(b) a decreased urinary excretion of calcium
(c) certain features which can be explained by excessive beta adrenergic stimulation
(d) a rise in the level of the plasma proteins which bind thyroxine.

480.

(a) T because of the metabolic acidosis, H^+ ions compete more effectively with K^+ ions for secretion by the renal tubules

(b) T due to the dissolved glucose

(c) F blood volume falls due to the osmotic diuresis

(d) F P_{CO_2} falls; the acidosis due to "ketone bodies" causes hyperventilation.

481.

(a) F it is negative due to muscle wasting (thyrotoxic myopathy)

(b) F it rises due to liberation of calcium salts from bone

(c) T beta adrenoceptor blockade alleviates some of the symptoms, e.g. tachycardia

(d) F the level of these proteins does not rise; they just bind more thyroxine.

482. The diagram shows the effect, in three individuals, of an intravenous injection of thyrotropin-releasing hormone (TRH) at time zero on blood levels of thyroid-stimulating hormone (TSH). Curve B is the response of a normal person. In this test:

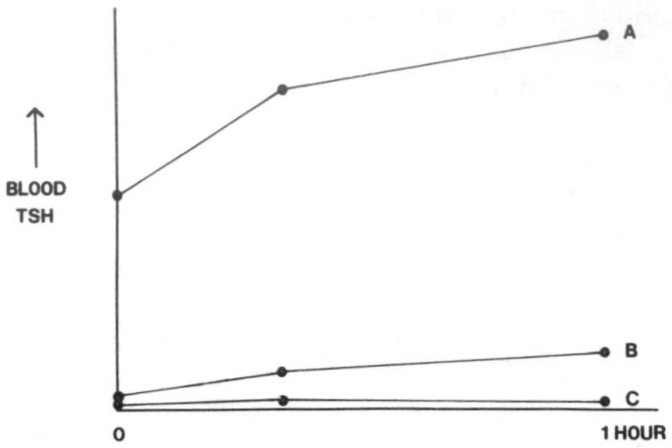

(a) the injected TRH is a protein composed of nearer 100 than 20 amino acids
(b) curve A rather than curve C is the response of a patient suffering from primary hyperthyroidism (i.e. not secondary to pituitary disease)
(c) the result is more discriminating from a diagnostic point of view in the case of primary hypothyroidism than for primary hyperthyroidism
(d) the result would be nearer curve B than curve A in the case of a patient with hypothyroidism secondary to pituitary insufficiency.

483. A patient with a secreting tumour of the adrenal medulla (phaeochromocytoma) typically has:
(a) a raised blood pressure which may be transient or constant
(b) a fine tremor of the extended hand
(c) a lowering of blood pressure when an alpha adrenoceptor antagonist is given
(d) an increase in the basal metabolic rate.

482.

(a) F it is a tripeptide and can be synthesised commercially

(b) F curve A is typical of hypothyroidism—loss of the negative feedback of thyroid hormones leads to increased pituitary secretion of TSH and increased pituitary responsiveness to TRH; curve C is typical of hyperthyroidism

(c) T separation from normal is much greater in the case of primary hypothyroidism

(d) T in this case, poor secretion of TSH is the cause of the hypothyroidism.

483.

(a) T due to intermittent or steady release of adrenaline and noradrenaline

(b) T adrenaline produces tremor; it is a beta effect

(c) T such a finding in a patient with hypertension suggests that a phaeochromocytoma is present

(d) T due to the stimulating effect of adrenaline on metabolism; the condition may mimic hyperthyroidism.

484. Full body stature is not attained by people who as children have experienced:

(a) chronic malnutrition

(b) premature puberty

(c) castration

(d) thyroid deficiency.

485. Insulin:

(a) requirements at night are approximately equal to those during the day

(b) usually has a shorter half-life than normal in patients with diabetes mellitus

(c) in the blood is partly bound to larger circulating proteins

(d) secretion may be increased by certain drugs.

486. The likelihood of developing tetany is increased when:

(a) plasma bicarbonate rises

(b) plasma magnesium rises

(c) the anterior pituitary is removed

(d) respiratory failure develops.

487. Destruction of the anterior lobe of the pituitary gland (Simmonds' disease) is characterised by:

(a) infertility

(b) a lowered ability to withstand severe illness or injury

(c) the passage of very large quantities of urine, e.g. 5–10 litres per day

(d) a decrease in the basal metabolic rate.

484.

(a) T the stunting of growth is not made good by adequate diet in later life

(b) T the sex hormones cause early closure of the epiphyses

(c) F due to delayed closure of the epiphyses, castrates tend to be somewhat taller than normal

(d) T adrenal and thyroid hormones are essential for normal growth.

485.

(a) F the requirements are much less at night; normally insulin is secreted almost entirely in response to meals

(b) F this disease is not usually due to rapid inactivation of insulin

(c) T abnormality of binding may be related to the causation of diabetes mellitus

(d) T these drugs stimulate the beta cells directly and may be used instead of insulin in the treatment of some diabetic patients.

486.

(a) T by raising the pH, this results in the plasma proteins binding more of the free calcium

(b) F magnesium ions like calcium ions tend to prevent tetany

(c) F the pituitary does not seem to be involved in calcium homeostasis

(d) F since this causes a respiratory acidosis, the protein binding of calcium is decreased.

487.

(a) T ovulation and menstruation cease due to absence of FSH and LH

(b) T due to loss of ACTH, glucocorticoid secretion cannot be increased in stress

(c) F ADH is released from the posterior pituitary

(d) T due to lack of TSH, thyroid activity is depressed.

488. Removal of the thyroid gland leads to:

(a) a high blood cholesterol level

(b) a prolongation of the reaction time for stretch reflexes
(c) a fine tremor of the fingers
(d) flattening of the oral glucose tolerance curve.

489. In diabetes mellitus:

(a) the osmolality of the extracellular fluid tends to fall

(b) the appetite is usually depressed
(c) the plasma bicarbonate level rises in severe cases
(d) the blood volume tends to fall in severe cases.

490. Excessive glucocorticoid production (Cushing's syndrome) is characterised by:

(a) thick skin

(b) demineralisation of bone
(c) hypertension
(d) slow healing of wounds.

491. Excess production of growth hormone by pituitary eosinophil cells may result in:

(a) giantism if it begins during childhood
(b) giantism if it begins during adulthood

(c) a raised blood glucose level
(d) enlargement of the liver.

492. Hypoglycaemic coma differs from hyperglycaemic coma (diabetic ketosis) in that there is:

(a) a more rapid loss of consciousness

(b) a weaker pulse

(c) an invariably negative urine test for glucose

(d) a greater chance of finding acetone in the urine.

488.
(a) T thyroxine normally stimulates the hepatic mechanisms which excrete cholesterol
(b) T the mechanism is obscure
(c) F this is a characteristic sign of excessive thyroid activity
(d) T due to lack of the normal facilitation of glucose absorption by thyroxine.

489.
(a) F it rises, due to the increased number of glucose particles; hence the characteristic thirst
(b) F the inability to use glucose for energy results in hunger
(c) F it falls; bicarbonate is used to buffer the keto acids
(d) T due to depletion of body water (especially extracellular) by osmotic diuresis and by vomiting.

490.
(a) F the skin is thin due to protein catabolism; purple striae appear on the abdomen
(b) T due to the breakdown of the protein matrix
(c) T due to both glucocorticoid and mineralocorticoid effects
(d) T glucocorticoids inhibit fibroblastic activity.

491.
(a) T due to an increased rate of general body growth
(b) F not after epiphyses have closed; it causes enlarged extremities (acromegaly) in adults
(c) T growth hormone has a diabetogenic action
(d) T body organs in general share in the excessive growth.

492.
(a) T the blood sugar level can drop more abruptly than the abnormal metabolites of diabetic ketosis can develop
(b) F the reverse is true due to fluid depletion in diabetic ketosis
(c) F sugar may have entered the bladder before the onset of the hypoglycaemia
(d) F the reverse is true.

493. In diabetic ketosis there is a decreased metabolic breakdown of:
(a) hepatic glycogen
(b) protein
(c) fat

(d) glucose.

494. Adrenal insufficiency causes:
(a) a fall in extracellular fluid volume

(b) increased breakdown of protein

(c) low blood pressure
(d) a rise in the plasma sodium: potassium ratio.

495. Appropriate treatment for a patient suffering from severe diabetic ketosis includes administration of:
(a) intravenous fluids
(b) intravenous insulin
(c) oxygen if hyperventilation is present

(d) lactate or bicarbonate if acidosis is marked.

493.
(a) F glycogen breakdown is increased
(b) F gluconeogenesis is increased
(c) F lipid catabolism is accelerated with increased production of ketone bodies
(d) T breakdown of protein and fat is increased to compensate for impaired glucose metabolism.

494.
(a) T this is secondary to the sodium loss resulting from lack of aldosterone
(b) F excessive glucocorticoid secretion increases protein breakdown
(c) T the low blood volume contributes to this
(d) F the ratio falls (aldosterone favours sodium retention and potassium excretion).

495.
(a) T this is essential to correct dehydration
(b) T this is also an urgent requirement
(c) F the hyperventilation is caused by acidosis; not oxygen lack
(d) T this is less important than (a) and (b).

496. The diagram shows results obtained during a glucose tolerance test on three people, of whom the person represented by curve B was normal. The oral glucose load was given at time zero. It can be deduced that:

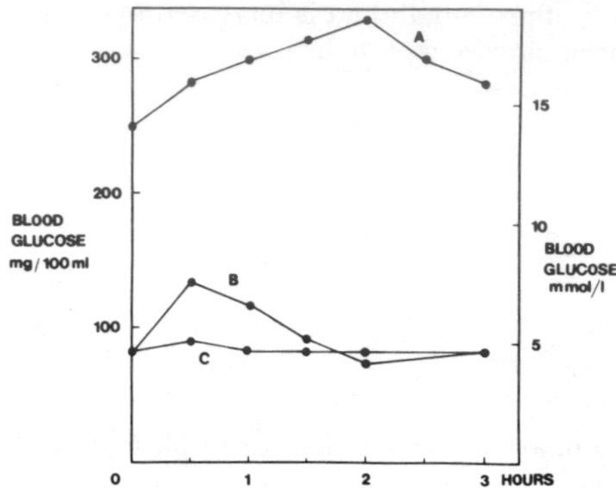

(a) curve A is consistent with a diagnosis of diabetes mellitus

(b) curve C is more consistent with a diagnosis of an insulin-secreting tumour than with a diagnosis of malabsorption

(c) a person whose blood glucose followed curve B and who had glucose in the urine 30 minutes after ingesting the glucose shows a low renal glucose threshold

(d) the renal clearance of glucose for the subject represented by curve A at 2 hours is nearer 10% than 30% of renal plasma flow, assuming a normal threshold.

497. Failure of the anterior pituitary to secrete growth hormone:

(a) results in reduction in body stature and organ size when it occurs in adults

(b) causes puberty to be delayed in children

(c) results in stunting of growth in children which is much more marked in the limbs than in the trunk

(d) can be treated effectively in children by injections of bovine growth hormone.

496.

(a) T the fasting glucose level and the peak level are markedly raised and there is delayed return of the blood glucose to the fasting level

(b) F curve C is typical of the flattening obtained with malabsorption; with an insulin-secreting tumour the fasting level would tend to be low, with a more marked rise to a peak, and a subsequent trough below the initial level

(c) T the normal renal glucose threshold is about 180 mg/100 ml (10 mmol/l)

(d) T about 150/330 of the filtered glucose is being lost, corresponding to a clearance of 40–50% of the glomerular filtration rate, i.e. about 60 ml which is about 10% of the renal plasma flow.

497.

(a) F it has no detectable effect on body size in adults

(b) F the release of the pituitary gonadotrophins are responsible for the timing of puberty

(c) F pituitary dwarfs are usually well proportioned; achondroplastic dwarfs have relatively short limbs

(d) F unfortunately, bovine growth hormone is antigenically distinct from human growth hormone; the human variety is needed for treatment.

498. Following surgical removal of the pituitary (hypophysectomy):

(a) insulin injections have a greater effect on the blood glucose level

(b) the blood level of aldosterone falls

(c) there is loss of pubic and axillary hair

(d) penile erection cannot occur.

499. Parathormone secretion is typically increased:

(a) in chronic renal failure

(b) following overdosage with vitamin D

(c) in patients immobilised for long periods in bed

(d) when the blood phosphate level falls.

98.

a) T growth hormone secretion will not be available to counteract the action of insulin

b) F the aldosterone level is regulated by the renin-angiotensin system

c) T due to deficiency of adrenal androgens which are under ACTH control

d) F penile erection depends on a spinal cord reflex; sexual desire (libido) is, however, diminished.

199.

(a) T phosphate retention tends to lower blood calcium which in turn increases parathyroid activity

(b) F the increase in blood calcium due to excessive vitamin D depresses the parathyroids

(c) F the demineralisation of bones that occurs in these patients is due to the reduction in mechanical stresses being applied to their bones; parathormone is not involved

(d) F the fall in phosphate would tend to increase the blood level of calcium and so reduce parathormone secretion.

REPRODUCTIVE SYSTEM
BASIC PHYSIOLOGY

500. The diagram below indicates a number of possible patterns of change of measurements made throughout pregnancy. Measurements are expressed as a percentage of the maximal value. In normal pregnancy, changes in maternal:

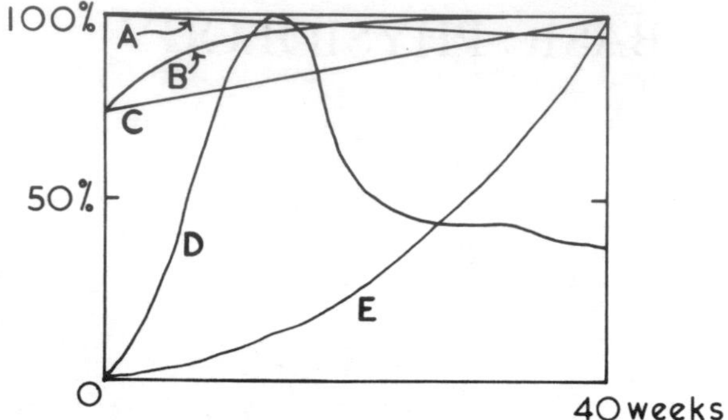

(a) urinary excretion of human chorionic gonadotrophin (HCG) correspond to line E rather than line D
(b) urinary oestrogen excretion correspond to line E rather than line C
(c) cardiac output follow line C rather than line B
(d) blood haemoglobin level follow line A rather than line C.

501. In the normal menstrual cycle:
(a) ovulation is associated with a sudden rise in blood LH levels
(b) the cervical mucus becomes more fluid about the time of ovulation
(c) the proliferative phase of endometrial growth depends on oestrogen secretion
(d) the blood loss during menstruation is about 300 ml.

500.

(a) F the reverse is the case, with maximal excretion of HCG in the first trimester

(b) T oestrogen excretion rises many-fold during pregnancy and provides some indication of placental function

(c) F the reverse is the case; much of the increase occurs in early pregnancy; uterine flow alone cannot account for this increase in early pregnancy, flow to other organs is also increased

(d) T increased haemoglobin synthesis tends to lag slightly behind the increase in blood volume so that the haematocrit falls.

501.

(a) T just before ovulation there is a surge of LH

(b) T this may facilitate the passage of sperms into the uterus; the "cascade" of mucus may be used to indicate the imminence of ovulation

(c) T the proliferative phase of endometrial growth occurs in the first half of the cycle and is due to oestrogens

(d) F though it varies considerably from person to person, 30 ml is about the average loss.

502. Fertilisation of the human ovum:

(a) normally occurs in the uterus

(b) by one sperm normally prevents other sperms from entering the ovum

(c) may occur up to 5 days after ovulation

(d) usually occurs about 4 days before implantation.

503. Human spermatozoa:

(a) are produced at a faster rate when testicular temperature is raised from 35 to 37°C

(b) are motile in the seminiferous tubules

(c) contain 23 chromosomes

(d) contain enzymes in their heads which facilitate penetration of the ovum.

504. After the fetus is born:

(a) its systemic vascular resistance rises

(b) its pulmonary vascular resistance rises

(c) there is reversal of flow in the ductus arteriosus

(d) an opening between the two ventricles closes.

502.
(a) F it normally occurs in the outer third of the uterine (Fallopian) tube
(b) T by an unknown mechanism it makes the shell (zona pellucida) impermeable to other sperms
(c) F the ovum remains viable for only about 24 hours
(d) T during this time the fertilised ovum is travelling down to the uterus.

503.
(a) F raising testicular temperature depresses sperm production; undescended testes are infertile
(b) F at this stage they are incapable of motility or fertilisation
(c) T half the complement of an ordinary human cell
(d) T these enzymes which are in the acrosome of the head are released when the sperm is in the female genital tract.

504.
(a) T due to the closure of the umbilical arteries
(b) F due to expansion of the lungs, it falls
(c) T the reversal of the pressure gradient between the pulmonary artery and aorta causes blood to flow from the aorta to the pulmonary artery
(d) F there is normally no interventricular opening; it is the interatrial foramen (foramen ovale) which closes.

505. In the newborn:
(a) the bilirubin level in the blood tends to be higher than that in the adult
(b) the brain cells are less tolerant to lack of oxygen than in the adult
(c) the ability to manufacture antibodies is poorly developed
(d) there is little resistance to infection.

506. Human chorionic gonadotrophic hormone (HCG):
(a) acts on the uterus to maintain the integrity of the endometrium in early pregnancy
(b) production is greatest in the last three months of pregnancy
(c) may be identified in the urine of pregnant women by an immunological technique
(d) is a steroid hormone.

507. Testosterone causes:
(a) depression of pituitary secretion of LH
(b) an increase in scalp hair
(c) the epiphyses of long bones to unite
(d) a negative nitrogen balance.

508. Androgens:
(a) are formed in the seminiferous tubules of the testis
(b) are formed in greater quantity in fetal life than in childhood
(c) are secreted in small amounts in adult females
(d) in the blood decrease in concentration after the age of 30 in men.

505.
(a) T this is due to the relative inefficiency of the liver in the neonate and is responsible for the "physiological jaundice" seen in the newborn; the liver does not produce enough glucuronyl transferase which conjugates bile pigments with glucuronic acid
(b) F the reverse is the case
(c) T in the fetus foreign proteins do not evoke an immune response. This ability develops about 3 months after birth
(d) F the baby has been supplied with maternal antibodies via the placenta and colostrum.

506.
(a) F it acts on the ovaries to maintain the corpus luteum and thus the output of progesterone
(b) F it is highest in the first three months; in the last six months of pregnancy the placenta and not the corpus luteum is responsible for progesterone production. The HCG may then act locally on the placenta so that a high systemic level is not necessary
(c) T this is the usual way of identifying the hormone in pregnancy tests
(d) F it is a glycoprotein; its structure is only slightly different from that of LH.

507.
(a) T this is a negative feedback phenomenon which normally maintains plasma testosterone at a nearly constant level
(b) F though it tends to increase body hair, scalp hair tends to recede
(c) T sexual precocity can cause short stature
(d) F the reverse is true; protein synthesis is increased and protein breakdown decreased; testosterone is an *anabolic* hormone.

508.
(a) F they are formed by the interstitial (Leydig) cells between the seminiferous tubules
(b) T secretion of androgens by the fetal testis is responsible for the development of male sex organs
(c) T by the ovary and adrenal glands
(d) T there is a gradual fall; the more gradual fall of the sex hormone blood levels in men as they age may explain why the male climacteric is less evident than that in the female.

509. During pregnancy:

(a) uterine muscle tissue is increased due to rapid multiplication of the smooth muscle cells
(b) spontaneous contractions of the uterine muscle do not occur, due to the action of progesterone
(c) the basal metabolic rate of the mother rises by about one-third
(d) the breasts enlarge due to the release of prolactin from the anterior pituitary gland.

510. In the mammary glands:

(a) milk formation is stimulated by oestrogen and progesterone
(b) maintenance of lactation depends on afferent impulses passing to the brain from sensory receptors in the nipple areas
(c) milk formation ceases if the anterior pituitary gland is destroyed
(d) milk ejection ceases if the posterior pituitary gland is destroyed.

511. Males differ from females in that:

(a) they have a greater tendency to have a "drumstick" of chromatin projecting from the nuclei of polymorphonuclear granulocytes
(b) their pituitary glands secrete different gonadotrophic hormones
(c) their gonadal gametogenic function persists until later in life
(d) they do not show a rise in gonadotrophic hormone levels in the second half of life.

512. The post-pubertal state differs from the pre-pubertal state in boys in that:

(a) the gonads are responsive to gonadotrophic hormones
(b) there is a greater output of 17-ketosteroids in the urine
(c) skeletal muscle is stronger per unit mass of tissue
(d) the blood contains a higher level of follicle stimulating hormone (FSH).

509.
(a) F the increase in bulk (about 15-fold) is due mainly to an increase in size of the individual cells
(b) F spontaneous (Braxton Hicks) contractions of the uterine muscle can be felt through the abdominal wall from the 12th week onwards
(c) T largely due to the intense metabolic activity in the fetus
(d) F breast enlargement is due to progesterone and oestrogen secretion. Prolactin is responsible for milk secretion after parturition.

510.
(a) F these tend to depress milk secretion but their blood levels fall dramatically after delivery of the baby
(b) T the initiation and maintenance of lactation after delivery depend on suckling of the nipples by the baby
(c) T lactation depends on release of prolactin from the anterior pituitary gland in response to suckling
(d) T the posterior pituitary hormone oxytocin is concerned in the expression of milk.

511.
(a) F females, with an XX chromatin pattern, show this pattern; males generally do not
(b) F pituitary gonadotrophins are the same in males and females
(c) T ovaries cease to function at about 45 years; testes cease to function much later
(d) F in both sexes gonadal function declines and levels of FSH and LH rise.

512.
(a) F the gonads are responsive to gonadotrophic hormones before puberty
(b) T sex hormones are converted to 17-ketosteroids in the liver and excreted in the urine
(c) T muscle strength increases rapidly after puberty perhaps because of the action of testosterone
(d) T FSH is responsible for the growth and function of the seminiferous tubules from the time of puberty.

513. In the placenta:

(a) there is mixing of maternal and fetal blood in the maternal sinusoids
(b) the oxygen pressure in the blood in maternal sinusoids is similar to that in alveolar air
(c) the blood leaving in the umbilical veins has an oxygen pressure similar (±5%) to that in the maternal sinusoids
(d) the blood leaving in the umbilical veins is more than 50% saturated with oxygen.

514. Fetal haemoglobin:

(a) has twice the iron content of adult haemoglobin
(b) is the only type of haemoglobin that can be identified in fetal blood
(c) has a higher oxygen capacity than adult haemoglobin
(d) forms the major fraction of total haemoglobin for the first year of life.

515. During pregnancy:

(a) the maternal blood volume doubles
(b) venous tone is decreased
(c) the placenta secretes a hormone which relaxes ligaments and connective tissue in the walls of the birth canal
(d) the output of maternal parathormone is increased.

516. The circulation in the fetus differs from that in the adult so that the blood in the:

(a) brachial artery has a higher oxygen content than that in the femoral artery
(b) superior vena cava has a higher oxygen content than that in the inferior vena cava
(c) right ventricle has a higher oxygen content than that in the left ventricle
(d) pulmonary artery has a higher oxygen content than that in the pulmonary veins.

513.

(a) F the two circulations remain discrete

(b) F it is about 40% of that in the lung alveoli

(c) F it is about 50% lower due to the barrier to diffusion offered by the placenta

(d) T because of the affinity of fetal blood for oxygen, a PO_2 of 20 mm Hg (2·6 kPa) is sufficient to give 70% saturation.

514.

(a) F it is in the globin part of the molecule that adult and fetal haemoglobin differ

(b) F adult haemoglobin starts to appear in mid-pregnancy and accounts for about 20% of the haemoglobin at the time of birth

(c) F but fetal blood takes up and releases most of its oxygen at lower oxygen tensions than does adult blood

(d) F by 4 months after birth fetal haemoglobin comprises only 10% of the total haemoglobin.

515.

(a) F it rises by about 30%

(b) T this may be a direct action of progesterone; it may contribute to the development of varicose veins

(c) T it is called *relaxin* and it depolymerises collagen fibres

(d) T this is thought to mobilise calcium for the fetus.

516.

(a) T deoxygenated blood enters the aorta through the ductus arteriosus below the origin of the arteries that supply the head and arms

(b) F oxygenated blood enters the inferior vena cava from the umbilical veins

(c) F deoxygenated S.V.C. blood tends to stream to the right ventricle whereas oxygenated I.V.C. blood tends to steam via the foramen ovale to the left ventricle

(d) T since the lungs are not ventilated, O_2 is lost rather than gained in a circuit through the fetal lungs.

517. Spermatozoa:

(a) when separated from the other constituents of semen are infertile

(b) are normally stored in the seminal vesicles before being expelled through the urethra

(c) require testosterone for normal development

(d) require FSH (follicle-stimulating hormone) for normal development.

518. Normal parturition is dependent on:

(a) the cessation of placental oestrogen and progesterone secretion

(b) the release of oxytocin from the posterior pituitary gland

(c) the presence of normally functioning ovaries

(d) an autonomic stretch reflex.

519. The size of the fetus at birth tends to be:

(a) related more to the size of the mother than to the size of the father

(b) less in the case of a twin, triplet, etc. than in the case of a single baby

(c) on average, greater in the female

(d) greater in first-born than in subsequent children.

520. Glandular cells which lie between the seminiferous tubules in the testis:

(a) secrete seminal fluid

(b) are stimulated to secrete by luteinising hormone (LH)

(c) are indirectly influenced by cells in the hypothalamus

(d) are unable to function unless the testis descends into the scrotum.

517.
(a) T fertility is conferred by mixing them with the other constituents
(b) F they are stored in the epididymis at a temperature below that of the body core
(c) T however, testosterone injections may abolish spermatogenesis by suppressing gonadotrophin secretion
(d) T hypophysectomy results in aspermatogenesis.

518.
(a) F secretion is maintained at high levels until parturition
(b) F people with diabetes insipidus due to loss of posterior pituitary function have had normal parturition
(c) F the ovaries are not essential for the latter stages of pregnancy or parturition
(d) F parturition can occur with a denervated uterus; the cause of the onset of labour is not known.

519.
(a) T as an extreme example, small Shetland mares inseminated with sperm from large Shire stallions do not have excessively large fetuses; this principle appears to apply in humans also
(b) T with multiple pregnancies the placental capacity per fetus is reduced
(c) F on average the male is about 200 g heavier
(d) F later children are on average about 200 g heavier than the first-born.

520.
(a) F these interstitial (Leydig) cells secrete testosterone
(b) T LH was originally called interstitial cell stimulating hormone in the male
(c) T these hypothalamic cells secrete LH/FSH releasing hormone
(d) F descent of the testis into the scrotum is necessary for spermatogenesis, not for testosterone production.

521. Testosterone:

(a) secretion shows a circadian rhythm, reaching a peak around 5–6 pm
(b) administration to healthy adults leads to an increase in the sperm count in their semen
(c) production increases at puberty because the level of circulating LH rises markedly
(d) secretion causes fat in subcutaneous tissue of the abdominal wall to be distributed mainly below the umbilicus.

522. The corpus luteum:

(a) plays an essential role in developing and maintaining the endometrium during the secretory phase of the cycle
(b) is under the control of the pituitary gland
(c) continues to secrete ovarian hormones in early pregnancy due to continued stimulation by pituitary gonadotrophins
(d) begins to atrophy around the end of the first month of pregnancy.

523. Erection of the penis:

(a) can occur in a patient whose spinal cord has been severed in the lower thoracic region
(b) does not occur before puberty
(c) is initiated by the constrictor action of noradrenaline on the veins draining the erectile tissue
(d) is essential for ejaculation.

521

(a) F peak testosterone blood levels are seen about 7–8 am

(b) F by decreasing the pituitary production of FSH, a raised level of blood testosterone reduces the sperm count

(c) T this is secondary to release of LH/FSHRH by the hypothalamus; LH stimulates the Leydig cells to produce testosterone which is responsible for the growth of the sexual organs and the development of the secondary sexual characteristics

(d) F the distribution is mainly above the umbilicus and in severe cases may overhang the belt! Oestrogens promote a sub-umbilical distribution.

522.

(a) T this phase is maintained by the progesterone secreted by the corpus luteum

(b) T LH from the pituitary is responsible for the development of the corpus luteum

(c) F in early pregnancy it is stimulated to secrete by placental human chorionic gonadotrophin (HCG)

(d) F it plays an essential role in the first three months of pregnancy.

523.

(a) T erection is based on a spinal reflex whose co-ordinating centre lies in the lumbosacral cord; the reflex is, of course, modified by higher centre activity

(b) F erection can occur in the absence of testosterone and is frequently seen in infants as a spontaneous occurrence

(c) F it is initiated by the vasodilator action of cholinergic parasympathetic nerves on the arterial supply to the erectile tissue so that it becomes distended by blood

(d) F ejaculation can occur with the penis in a flaccid state but it is difficult to introduce into the vagina in this condition.

524. The placenta:
(a) contains villi whose cells transport glucose into the fetal blood by facilitated diffusion

(b) can convert glucose into glycogen and store it

(c) actively transports oxygen from maternal to fetal blood

(d) allows protein molecules to pass from the maternal to the fetal blood by the process of pinocytosis.

525. The ovaries:
(a) start to form ova at puberty

(b) discharge 5–10 ova during each menstrual cycle
(c) may be influenced by hormones from the placenta

(d) are essential for cyclical uterine activity.

526. The normal seminal ejaculation:
(a) has an average volume of about 2–5 ml

(b) comes mainly from the seminiferous tubules and epididymis

(c) contains buffers which make the pH of the vaginal fluids more suitable for sperm viability
(d) contains prostaglandins.

524.

(a) T though the fetal blood glucose is at a lower concentration than maternal blood glucose, glucose uptake by the fetus is greater than can be accounted for by simple diffusion

(b) T the placenta can also store proteins, fats, iron and calcium

(c) F oxygen movements can be accounted for by the prevailing oxygen concentration gradients

(d) T gamma globulins are transported in this way and are responsible for the passive immunity conferred by the mother on the fetus in late pregnancy.

525.

(a) F they contain 250 000–500 000 immature ova at birth; no more are formed after this

(b) F rarely do they discharge more than one per cycle

(c) T the developing placenta produces chorionic gonadotrophic hormone which maintains the activity of the corpus luteum in early pregnancy

(d) T because of their hormonal secretions.

526.

(a) T it has an average spermatozoa content of about 10^8 per ml

(b) F these contribute only about 20%; fluid from the seminal vesicles comprises about 60% of the volume

(c) T the buffers tend to neutralise the acidity of the vaginal fluids

(d) T prostaglandins were first discovered in semen; they stimulate smooth muscle in many situations but their function is obscure.

527. Compared with the 7th day of the menstrual cycle, on the 21st day there is a greater:

(a) progesterone level in the blood

(b) thickness of uterine muscle

(c) body temperature

(d) thickness of endometrium.

528. Compared with that of a normal adult, the newborn infant's:

(a) kidneys have less ability to excrete a concentrated urine

(b) blood-brain barrier is less permeable to bilirubin

(c) heat regulation is more efficient because of its ability to metabolise brown fat

(d) blood has a greater affinity for oxygen at low oxygen pressures.

529. Changes in maternal physiology during pregnancy include:

(a) a positive nitrogen balance

(b) a moderate rise in the diastolic blood pressure, e.g. from 80 to 100 mm Hg (10·6 to 13·3 kPa) in the latter half of pregnancy

(c) a relaxation of smooth muscle in the alimentary tract and ureters

(d) a rise in the renal threshold for glucose.

527.

(a) T in the second half of the cycle, LH secretion by the pituitary maintains the corpus luteum which in turn secretes progesterone

(b) F there is no appreciable change in the amount of uterine muscle throughout the menstrual cycle

(c) T the rise in body temperature is an indicator that ovulation has occurred

(d) T due to development of the mucosal glands.

528.

(a) T the adult kidney can concentrate urine to about 4 times and the neonatal kidney to about twice the osmolality of plasma

(b) F the reverse is true; hence high levels of bilirubin may damage the infant's brain

(c) F despite this ability, the infant can cope less well with large changes in environmental temperature

(d) T the oxygen dissociation curve for fetal and neonatal blood is shifted to the left of that for adult blood.

529.

(a) T about 300 g nitrogen are retained during the average pregnancy, half of which is retained by the mother and half by the fetus

(b) F blood pressure tends to fall; a rise such as that quoted is seriously abnormal

(c) T this may cause heartburn and dilatation of the ureters

(d) F the renal threshold falls; glucose may spill over in the urine with moderately low blood glucose levels.

530. In the diagram below which shows the changes in ovarian hormone concentration (different scales apply to A and B) in the blood throughout the menstrual cycle:

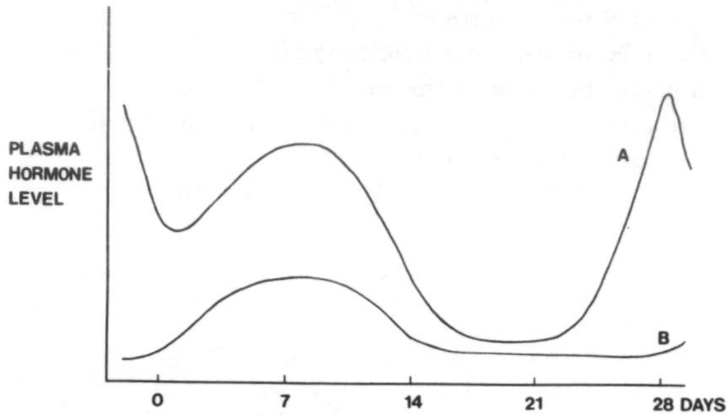

(a) the onset of menstruation could appropriately be indicated around day 0

(b) hormone B is oestrogen rather than progesterone

(c) the corpus luteum is responsible for the blood levels of both hormones at day 7

(d) day 21 would lie in the secretory rather than the proliferative phase of the endometrial cycle.

531. Amniotic fluid is:

(a) produced mainly by the fetal kidneys in late pregnancy

(b) actively swallowed by the fetus

(c) similar in its protein and electrolyte content to plasma

(d) inhaled and exhaled by the fetus.

530.

(a) F menstruation would commence about day 14 following the fall in the blood concentration of both ovarian hormones
(b) F it is progesterone which reaches its peak concentration about 7 days before menstruation
(c) T the corpus luteum secretes both progesterone and oestrogen
(d) F the reverse is true.

531.
(a) T in early pregnancy before the skin is keratinised it is formed mainly by filtration from the fetal skin capillaries
(b) T the amount swallowed, which may be as much as half a litre per day at term, is absorbed by the gut and excreted by the kidneys
(c) F the electrolyte concentrations are similar to those in plasma but the protein content is much lower
(d) T respiratory movements are seen quite early in pregnancy. Surfactant is laid down in the fetal lungs towards the end of pregnancy and its presence in amniotic fluid is used as an index of fetal maturity.

532. In the days following ovulation:

(a) the plasma progesterone level rises
(b) the cervical mucus becomes more viscous and scanty

(c) the plasma level of luteinising hormone falls

(d) the body temperature falls.

533. Ejaculation of semen:

(a) involves rhythmic contractions of striated muscles

(b) is aided by sympathetic nerve activity

(c) is accompanied by contraction of the cremasteric muscles
(d) is followed by orgasm.

534. In the fetal blood leaving the maternal sinusoids in the placenta:

(a) the glucose level is less than that in the maternal blood

(b) the hydrogen ion concentration equals that in the maternal blood

(c) the haemoglobin concentration is higher than in maternal blood

(d) may be found maternal pituitary hormones if these are being secreted by the mother.

535. The fetus normally:

(a) gains more weight in the last 10 weeks of gestation than in the first 30 weeks
(b) stores sufficient iron in the liver so that it does not require iron in the diet for the first year after birth
(c) has roughly the same basal metabolic rate as the adult when it is expressed per unit area of body surface

(d) passes some meconium into the amniotic fluid during the last three months of pregnancy.

532.
(a) T it is formed by the developing corpus luteum
(b) T this effect is due to the action of progesterone and makes the cervix less penetrable to sperms
(c) T it reaches its peak concentration in the cycle just before ovulation
(d) F the body temperature rises by about 1°C in the second half of the cycle.

533.
(a) T the ischiocavernosus and bulbospongiosus muscles cause rhymical compression of the urethra
(b) T this constricts the internal sphincter of the bladder and prevents reflux of semen into the bladder
(c) T this causes elevation of the testicles
(d) F ejaculation and orgasm coincide.

534.
(a) T this is so, despite the fact that glucose is transported into the fetal blood by facilitated diffusion
(b) F because of the diffusion barrier offered by the placenta fetal pH does not equalise with maternal pH by the time it leaves the placenta
(c) T about 17–20 g% in the fetus compared with about 12–14 g% in the mother
(d) F these hormones do not cross the placenta; transfer of proteins, e.g. antibodies, is a highly selective process of pinocytosis.

535.
(a) T fetal growth is exponential
(b) F iron is stored in the liver, but only enough for a few months after birth
(c) F because of rapid growth, the fetal metabolic rate is about double that of the adult when corrected for differences in body surface area
(d) F the passage of meconium into the amniotic fluid is a sign of fetal distress.

536. The fetal:

(a) blood leaving the placenta contains amino acids at a higher concentration than does maternal blood entering the placenta

(b) blood pressure in the aorta is lower than that in the pulmonary artery

(c) heart rate suggests fetal distress if it rises above 100 beats/minute

(d) blood glucose level is related to the rate of maternal uterine blood flow.

537. The neonatal:

(a) liver stores sufficient vitamin K for the first few months of life

(b) digestive tract lacks some of the enzymes needed for the adequate digestion of milk protein

(c) blood glucose level is liable to fluctuate more than the fetal blood glucose level

(d) blood volume is closer to 750 than 250 ml.

538. After birth:

(a) the normal baby's haemoglobin concentration would be expected to rise by about 20% in the first year

(b) the umbilical cord should be clamped as soon as possible to prevent blood loss from the fetus

(c) a normal child would be expected to double his birth weight rather than treble it in a year

(d) the baby's blood glucose level is normally lower than that found in older children and adults.

536.
(a) T because amino acids are actively transported from the maternal blood by the placenta into fetal blood
(b) T this results from the relatively low systemic resistance and the relatively high pulmonary resistance and causes blood to pass from the pulmonary artery to the aorta through the ductus arteriosus
(c) F the normal fetal heart rate is about 140 beats/minute; a rate of less than 100 suggests fetal distress
(d) T the uterine blood flow supplies the placenta; all other things being equal, the greater the blood flow, the higher the fetal blood glucose level.

537.
(a) F vitamin K is not stored and synthetic vitamin K is often given either to the mother just before delivery or to the newborn infant to reduce the danger of fetal bleeding. The absence of an intestinal flora precludes intestinal synthesis of vitamin K
(b) F the enzymes of the neonatal digestive tract are quite capable of dealing with the fat, carbohydrate and protein that are found in milk
(c) T the neonatal liver is less able to correct changes in blood glucose than is the mother's liver
(d) F it is closer to 250 ml (approximately 80 ml/kg as in the adult).

538.
(a) F it falls from around 20 g/100 ml at birth to around 11 g/100 ml at one year; this results more from the breakdown of cells containing fetal haemoglobin and from the immaturity of the bone marrow than from lack of iron
(b) F the placenta contains about half as much blood as the fetus and some of this blood should be transferred to the fetus at birth by uterine contraction, etc. Clamping of the cord should be delayed by not less than 20 seconds after delivery of the fetus.
(c) F birth weight is usually doubled between 3 and 6 months and trebled at a year
(d) T this may be due to the immaturity of the fetal liver; the baby is not considered to be hypoglycaemic unless the level falls to below 1·1 mmol/l (20 mg/100 ml).

REPRODUCTIVE SYSTEM APPLIED PHYSIOLOGY

539. Failure of a testis to descend:

(a) is associated with failure to produce testosterone

(b) does not interfere with fertility

(c) should be treated by injections of testosterone

(d) makes it more prone to malignant change.

540. Lack of surfactant in the lung:

(a) tends to be more marked the more prematurely is the infant born

(b) increases the muscular effort the infant has to make during expiration

(c) leads to poor oxygenation of the blood before birth

(d) in the infant may result in blood being shunted from the right to the left atrium through the foramen ovale and from the pulmonary artery to the aorta through the ductus arteriosus.

541. A premature infant is more likely than a full-term infant to:

(a) suffer from jaundice of hepatic origin

(b) maintain a normal body temperature in a cold environment (e.g. in a cot in a room at 5–10°C)

(c) excrete urine with a uniform specific gravity

(d) suffer from anaemia.

539.
(a) F the Leydig cells function normally when the testes are in the abdomen
(b) F spermatogenesis is seriously defective because of the high testicular temperature
(c) F this might cause premature closure of the epiphyses, and dwarfism; the testes should be brought from the abdominal cavity to the scrotum and fixed there by surgical means
(d) T for this and the above reason (b), surgical correction is best carried out in childhood.

540.
(a) T surfactant is first formed in the lung alveoli from about the 35th week of pregnancy. It is a lecithin and fetal maturity can be estimated from its concentration in amniotic fluid
(b) F a greater muscular effort is needed in inspiration; with lack of surfactant, the surface tension of alveolar fluid is so high that the lungs tend to collapse and are hard to inflate
(c) F it affects oxygenation only when the lungs are required for gas exchange
(d) T the alveolar hypoxia caused by poor pulmonary ventilation may cause pulmonary vasoconstriction and pulmonary hypertension so that pressure gradients are established that cause right-to-left shunts.

541.
(a) T there is impaired ability to conjugate bilirubin because of inability of the immature liver to synthesise glucoronyl transferase
(b) F heat regulating mechanisms mature in late pregnancy and early infancy; premature infants have a high surface area to volume ratio and are nursed in incubators
(c) T the immature kidney is less able to concentrate or dilute the urine; fluid balance must be carefully maintained
(d) T iron stores are laid down in the fetal liver towards the end of pregnancy.

542. A child whose sex chromosome pattern is:
(a) XY develops into a normal female
(b) XO shows incomplete sexual maturation at puberty
(c) XXX develops excessive female secondary sexual characteristics
(d) XXY develops into a true hermaphrodite.

543. Development of the secondary sexual characteristics before the age of 9 years:
(a) may be caused by inappropriate secretion of adrenal cortical hormones
(b) is not due to any associated serious disease in the majority of cases
(c) is associated with short stature
(d) may be caused by a tumour in the hypothalamus.

544. Temporary cessation of menstruation (secondary amenorrhoea):
(a) may occur for psychological reasons
(b) occurs if the body weight falls below a critical level
(c) may be caused by continuous administration of oestrogens or progestogens or both
(d) may occur in adrenal hyperplasia.

545. A diagnosis of pregnancy is suggested by:
(a) the presence in the urine of pregnanediol, a conjugated form of progesterone
(b) the finding that the cervical mucus becomes very viscous and forms a plug filling the cervical canal
(c) enlargement of the sebaceous glands in the mammary areola
(d) a hardening of the cervical tissue due to contraction of the smooth muscle around the mouth of the uterus.

542.
(a) F he develops into a normal male
(b) T the gonads are rudimentary or absent (Turner's syndrome)
(c) F no abnormalities seem to develop
(d) F he develops as a male with abnormal seminiferous tubules and a high incidence of mental retardation (Klinefelter's syndrome). True hermaphrodites probably have an XX/XY pattern.

543.
(a) T tumours of the adrenal may secrete sufficient androgens to cause sexual development in young children
(b) T remember the wide scatter associated with the normal distribution
(c) T the sex hormones cause early closure of the epiphyses
(d) T if the tumour cells secrete a hormone which stimulates the pituitary gland to secrete gonadotrophins.

544.
(a) T not uncommonly, factors such as severe depression, a change in environment, or marital disharmony may depress the secretion of gonadotrophin-releasing hormones from the hypothalamus
(b) T secondary amenorrhoea is one of the features of anorexia nervosa and starvation
(c) T it is withdrawal of these hormones which causes menstruation
(d) T androgens released by the adrenals oppose the effects of oestrogens on the endometrium.

545.
(a) F pregnanediol is always found in the urine of women during the child-bearing years; human chorionic gonadotrophin (HCG) in the urine indicates pregnancy
(b) T progesterone increases the viscosity of cervical mucus
(c) T these are known as Montgomery's tubercles
(d) F the cervix softens due to an increase in interstitial fluid and vascularity in the tissue.

546. Removal of the testes in the adult tends to cause:

(a) changes in the pitch of the voice

(b) loss of libido but not the ability to copulate

(c) hot flushes, irritability and depression

(d) a fall in the blood levels of LH and FSH.

547. Puberty does not occur in:

(a) children who have been castrated

(b) dwarfs

(c) boys whose testicles have been exposed to ionising radiation which has severely damaged the seminiferous tubules but not the interstitial (Leydig) cells

(d) children suffering from severe malnutrition.

548. Complications of using contraceptive pills based on oestrogen analogues include:

(a) a worsening more commonly than an improvement of acne vulgaris

(b) impaired haemostasis more commonly than a tendency to thrombosis

(c) sodium retention more commonly than sodium loss

(d) impaired liver function.

546.
(a) F though many secondary sexual characteristics regress, the voice does not change
(b) T testosterone secretion increases libido but is not essential for the reflex mechanisms needed for erection and ejaculation
(c) T rather like menopausal symptoms
(d) F the levels rise due to loss of feedback inhibition by testosterone.

547.
(a) T the hormones which produce the changes in puberty are produced in the gonads
(b) F it depends on the cause of the dwarfism; puberty may not occur in pituitary dwarfs
(c) F sterility results from damage to the seminiferous tubules. The Leydig cells are responsible for testosterone secretion and hence for the onset of puberty
(d) F puberty is delayed but usually occurs.

548.
(a) F oestrogens, by antagonising the effect of androgens on the sebaceous glands in the skin, tend to improve acne
(b) F on the contrary, there is an increased tendency for the blood to clot, so causing thrombosis at various sites
(c) T this is one of the effects of oestrogens; the associated retention of chloride and water expands the extracellular fluid compartment and causes weight gain
(d) T though oestrogens do not damage liver cells they reduce, by an unknown mechanism, the ability of the liver to excrete certain substances; mild jaundice may occur.

549. The diagram below indicates the pattern of change over some weeks of a woman's oral temperature taken before rising in the morning. The pattern:

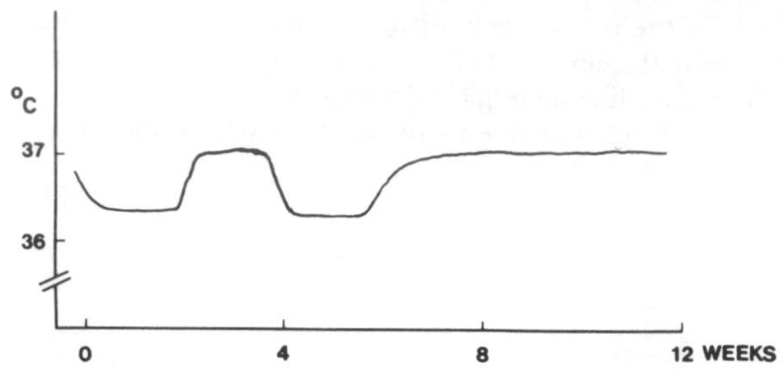

(a) suggests failure of ovulation between weeks 4 and 8

(b) is related to a depressant effect on basal body temperature of progesterone

(c) suggests a greater likelihood of there being a living ovum in the uterine tube around time 0 than at 2 weeks

(d) suggests a greater likelihood of there being a fertilised ovum in the uterus at week 9 than at week 5.

550. Oral contraceptive treatment consisting of mixtures of oestrogens and progestogens:

(a) if given daily throughout the year would tend to prevent menstruation from occurring

(b) is thought to act mainly by preventing implantation of the fertilised ovum

(c) is thought to depress anterior pituitary secretion of gonadotropic hormones

(d) may cause an increase in body weight in some patients.

549.

(a) F the rise in basal body temperature around week 6 suggests ovulation at that time

(b) F the body temperature tends to rise as the progesterone level rises after ovulation

(c) F the ovum is most likely to have entered the uterine tube around week 2; it generally survives for only one or two days

(d) T maintenance of the rise in body temperature from week 8 suggests that fertilisation and implantation have occurred.

550.

(a) T the drugs are usually given for 3 weeks out of 4; withdrawal leads to menstruation

(b) F it is believed to act by suppressing ovulation

(c) T by a negative feedback mechanism; thus the stimulus for ovulation is removed

(d) T sodium retention is one of the effects of these steroid drugs, perhaps because of some structural affinity with mineralocorticoids.

551. Methods of reducing fertility include:

(a) bilateral tubal ligation or vasectomy

(b) confining intercourse to the period from the 10th to the 20th day of the menstrual cycle

(c) the use of agents which prevent the fertilised ovum from implanting successfully in the uterine mucosa

(d) mechanical methods (condoms and caps), these being the most effective methods in routine use.

552. Infertility is usually:

(a) present when the sperm count in the ejaculate is around 10^7/ml (10% of normal)

(b) present when function in the posterior pituitary in the female is lost

(c) due to a defect of function in the female partner rather than the male partner

(d) due to a disorder of endocrine function in either the male or female partner.

553. Maternal blood loss from the genital tract in the first 24 hours after delivery:

(a) is considered abnormally great if it exceeds 600 ml

(b) tends to be greater after a short labour than after a long labour

(c) tends to be increased by failure to expel part of the placenta

(d) may, if excessive, affect the mother's anterior pituitary function.

551.
(a) T in the latter case an interval is necessary before infertility can be assumed
(b) F this is the period of maximum fertility; the "safe period" is the rest of the cycle
(c) T an intrauterine device (IUD) is thought to produce endometrical changes which are inimical to implantation
(d) F the oral contraceptives are the most effective, followed by the intra-uterine device.

552.
(a) T this seems surprising, since only one sperm ultimately fuses with the ovum; normal ejaculate contains about 10^8 sperms/ml
(b) F vasopressin and oxytocin are not essential for fertilisation
(c) F a functional defect in the male partner is about equally as common a cause of infertility as a defect in the female partner
(d) F the majority of cases of infertility are due to disorders affecting the reproductive tract

553.
(a) T is then called *primary post-partum haemorrhage*
(b) F prolonged labour is associated with post-partum haemorrhage, perhaps because the exhausted uterine muscle is less able to contract firmly and close the maternal sinusoids by compression
(c) T this would also limit the ability of the uterus to contract down and compress the bleeding vessels
(d) T the death of the anterior pituitary cells may result from ischaemia of the gland during the haemorrhagic shock. The condition is called Sheehan's disease and it is a recognised complication of post-partum haemorrhage.

554. Failure to ovulate would be considered the likely cause of female infertility if:

(a) the cervical mucus in the second half of the menstrual cycle showed "ferning" or "arborisation" when dried on a glass slide

(b) pregnanediol appeared in the urine in the second half of the cycle

(c) endometrial biopsy showed that the uterine mucosa contained glands in the second half of the cycle

(d) vaginal cytology showed large numbers of intermediate cells folded and aggregated into clumps.

555. The physical state of a baby in the first few minutes after a difficult delivery is worse if:

(a) its colour is blue rather than pale grey
(b) the heart rate is falling rather than rising

(c) there are spontaneous movements of the limbs rather than no movement

(d) it responds to a catheter in the pharynx with a grimace rather than a cough.

556. In premature labour, as opposed to late labour, there is a greater:

(a) risk of maternal complications

(b) risk of cerebral haemorrhage in the fetus because of the smaller size of the fetal head
(c) chance that the fetus, when born, has a pale skin

(d) fetal head to body-size ratio.

554.

(a) T the "ferning", which is due to the crystallisation of sodium chloride, is due to the unopposed action of oestrogen and is antagonised by progesterone. Its persistence would suggest that ovulation had not occurred, since a luteal phase had not commenced

(b) F pregnanediol is a degradation product of progesterone and its excretion in the urine would suggest a post-ovulatory phase

(c) F gland structures are present before and after ovulation. After ovulation they become more coiled and begin to secrete

(d) F this is typical of the luteal phase of the cycle and suggests that ovulation has occurred.

555.

(a) F a pale grey colour suggests that the circulation is failing

(b) T a rising heart rate is a good sign; it suggests that the baby is recovering from the vagal overactivity induced by hypoxia during delivery. A slowing heart rate suggests the reverse

(c) F after severe hypoxia the limb muscles are limp and reflexes are absent. Increasing muscle tone and movements are good signs

(d) T this indicates that the reflexes are depressed; this and the characteristics mentioned above can be scored so that the physical state of the baby is roughly quantitated, e.g. by the Apgar system of assessment.

556.

(a) F delivery is easier since the fetus is smaller and the mechanism of labour is not affected

(b) F cerebral haemorrhage *is* more common but the cause is the greater fragility of the skull and intracerebral vessels

(c) F the skin is brick red largely due to the relative lack of subcutaneous fat

(d) T in late pregnancy the body grows relatively faster than the head.

557. Following the female climacteric, when the level of ovarian hormones falls:

(a) vaginal pH falls

(b) the myometrium decreases in size

(c) the level of pituitary gonadotrophins in the blood falls

(d) there may be an increase in libido.

558. Women having their first child after the age of 35 differ from younger women having their first child in that they have a lower incidence of:

(a) excessive blood loss within 24 hours of delivery (post-partum haemorrhage)

(b) fetal abnormalities

(c) abortion

(d) delivery by Caesarian section.

559. Infertility in the male can be:

(a) explained by failure of the sperms in a sample of semen to remain motile for more than 24 hours

(b) explained by finding that 50% of the sperms in a sample are definitely abnormal

(c) explained by finding a sperm count which is 50% below average

(d) treated effectively in some cases with large doses of vitamin E.

557.

(a) F it rises; the decrease in acidity permits pathogenic organisms to multiply more readily

(b) T some of the muscle cells are replaced by fibrous tissue and the whole uterus decreases in size

(c) F it rises; the climacteric is essentially an ovarian failure and the lack of ovarian hormones removes the feed-back suppression of gonadotrophin secretion

(d) T libido may increase or decrease; most commonly there is little change.

558.

(a) F with increasing age the uterine contractions after labour are less effective in stopping bleeding

(b) F ova are formed in mid-fetal life and gradually die off from that time on; elderly ova seem more likely to lead to fetal abnormality, especially mongolism

(c) F abortion is two to three times more common, probably because of the higher incidence of fetal abnormality

(d) F normal delivery is less easy for the elderly primigravida since her perineum and vagina are less compliant and her uterine action is less efficient than those of a younger woman.

559.

(a) F sperms retain motility *in vitro* for only a few hours; loss of motility in less than one hour, however, would be considered abnormal

(b) T the other 50% which *appear* normal are unlikely to fertilise an ovum

(c) F the count has to be less than 20% of normal to cause infertility, provided the sperms themselves are normal

(d) F there is no agreement that vitamin E improves human fertility.

560. Failure of the newborn child to breathe adequately:
(a) may be due to sedative drugs given to the mother during labour
(b) results in muscle rigidity
(c) may be due to the fact that the respiratory tract is blocked with amniotic fluid and meconium
(d) should not be treated by giving the fetus oxygen to breath because of the risk of causing retrolental fibroplasia in the eyes.

561. Pregnant women who have had five or more previous babies differ from those who have had none in that they have a greater risk of:
(a) developing anaemia
(b) involuntarily passing small amounts of urine while coughing or laughing (stress incontinence)
(c) their baby being presented to the pelvis in an unusual position (malpresentation)
(d) complications due to Rhesus incompatibility.

562. Secretion of androgens in the adult female:
(a) does not normally occur
(b) by an androgen-secreting tumour may result in enlargement of the clitoris
(c) by an androgen-secreting tumour causes no change in the pitch of the voice
(d) by an androgen-secreting tumour may result in amenorrhoea.

563. Fetal death is the typical result of absence of fetal:
(a) liver function
(b) kidney function
(c) cardiac function
(d) brain function.

560.
(a) T many sedative and analgesic drugs depress respiration and most drugs cross the placenta readily
(b) F such babies become pale and flabby as muscle tone disappears and circulatory failure develops
(c) T sucking out the airways of these babies is routine treatment
(d) F oxygen should be given through a face mask with gentle intermittent positive pressure to inflate the lungs; retrolental fibroplasia is seen only in premature infants who breathe high concentrations of oxygen for long periods.

561.
(a) T this may be attributed to the depletion of maternal iron stores by the multiple pregnancies and the poorer diet that is frequently their lot
(b) T damage to the pelvic floor in child-birth may diminish the effectiveness of the sphincteric system of the bladder and so cause stress incontinence
(c) T the flabby uterine and abdominal wall muscles allow the baby's position to change easily
(d) T due to sensitisation to Rhesus antigen in earlier pregnancies.

562.
(a) F androgens are normally secreted by the adrenal cortex and cause growth of axillary and pubic hair
(b) T it may grow to resemble a small penis
(c) F the voice deepens due to enlargement of the larynx
(d) T since androgens antagonise the effects of oestrogen on the endometrium.

563.
(a) F the maternal liver can cover for the failure of the fetal liver
(b) F again the mother can cover for the fetal kidneys
(c) T the mother's circulation cannot substitute for that in the fetus
(d) F a brain is not a necessity for fetal survival.

564. A diagnosis of fetal distress can appropriately be made if:
(a) the fetus keeps making vigorous repetitive movements
(b) blood taken from a fetal scalp vein has a pH of about 7·25

(c) meconium is present in the amniotic fluid

(d) there is a sustained increase in the fetal heart rate.

565. Deficient secretion of testosterone results:
(a) in a reduction in male fertility

(b) in a paler complexion

(c) where there is a genetic disorder in which the male carries an extra X chromosome

(d) in high levels of LH in the blood in most cases.

566. Malformation of the fetus:
(a) can of itself cause abortion in the 8–12 week period of pregnancy

(b) such as imperforate ureters can cause acidosis in the fetus due to renal failure

(c) such as spina bifida can be diagnosed in some cases by changes in the composition of the amniotic fluid

(d) such as a connexion between oesophagus and trachea may cause no trouble until the fetus is born.

564.
(a) F vigorous fetal movements are normal and common
(b) F this is normal for the fetus; 7·1 would indicate a moderate to severe acidosis
(c) T when the fetal ability to deliver oxygen to its tissues decreases severely, there is increased peristaltic activity in the alimentary tract which drives the green intestinal contents (meconium) into the amniotic sac
(d) F fetal hypoxia tends to cause slowing of the fetal heart; the normal rate is about 140 beats/min and a fall to below 100 beats/min suggests distress.

565.
(a) T testosterone is needed for the proper maturation of sperms
(b) T testosterone normally causes increased vascularity, thickness and ruddiness of the skin together with increased growth and function of sebaceous glands
(c) T the testes fail to develop properly in Klinefelter's syndrome where there is a chromosome make-up of 44 plus XXY
(d) T except where the low testosterone level is due to failure of LH production by the pituitary.

566.
(a) T all the rudimentary organs are formed at this stage and if the malformation is incompatible with fetal life, abortion occurs; such malformations are the commonest cause of abortion at this stage
(b) F the maternal kidneys carry out all the renal functions the fetus needs; however, the volume of amniotic fluid may be reduced in late pregnancy, since it is derived mainly from the fetal kidneys at this stage
(c) T in some cases proteins derived from the central nervous system may pass into amniotic fluid through the spinal opening
(d) T since the mother acts as the fetus's lungs, kidneys, liver, alimentary tract and skin, defects of function in these organs will cause trouble only after birth when the fetus is forced to use its own organs.

GENERAL QUESTIONS
BASIC PHYSIOLOGY

567. Ultrafiltration is:

(a) a process in which the colloidal contents of a solution are separeted from the crystalloid contents
(b) entirely responsible for the formation of cerebrospinal fluid (CSF)
(c) entirely responsible for the formation of the glomerular filtrate
(d) responsible for the formation of tissue fluid by the capillaries.

568. The Fick principle enables:

(a) flow, F, to be calculated from the formula $F = U. (A - V)$, where U is the uptake or output of the substance by the organ in unit time and A and V are its mean concentrations in blood entering and leaving the organ
(b) cardiac output to be estimated using the lungs as the organ and carbon dioxide as the substance eliminated
(c) renal plasma flow to be calculated from the renal output of PAH divided by the concentration of PAH in arterial blood
(d) brain blood flow to be calculated from measurements of radioactivity over the skull as a bolus of radioactive material passes through the cerebral circulation.

569. Mitochondria:

(a) are oval-shaped membranous intracellular organelles containing cristae or shelf-like structures
(b) are concerned with energy production and its storage as ATP
(c) contain RNA and are the chief site for protein synthesis
(d) are more numerous in cells with high than in those with low energy expenditure.

570. Sleep is:

(a) associated with a well-marked alpha rhythm in the electroencephalogram
(b) hindered by a high level of circulating catecholamines
(c) not associated with any appreciable change in the blood pressure
(d) associated with a rise in central body temperature.

567.
(a) T this is a definition of ultrafiltration

(b) F active secretion is involved in the formation of CSF

(c) T there is no evidence that the glomerular cells can secrete fluid into Bownman's capsule

(d) T the process is not identical in all capillaries; skeletal muscle capillaries are much less permeable to protein than liver capillaries.

568.
(a) F the formula is $F = U/(A - V)$, since uptake must equal flow multiplied by the arteriovenous difference

(b) T U is the carbon dioxide output from the lungs, A and V are taken as the concentrations of carbon dioxide in pulmonary trunk and systematic arterial blood

(c) T since PAH at low concentrations is completely eliminated from the blood passing through the kidney, its renal venous concentration is zero. Therefore $F = U/A$

(d) F this method does not measure uptake and does not use the Fick principle.

569.
(a) T as seen by electron microscopy

(b) T mitochondria contain the enzymes used in the citric acid cycle; they are concerned with oxidative phosphorylation

(c) F this applies to the ribosomes

(d) T in fact they are most numerous in those regions of cells where physiological mechanisms are concentrated, e.g. close to the membranes of actively secreting cells.

570.
(a) F the alpha rhythm disappears and may be replaced by a delta rhythm

(b) T catecholamines tend to increase alertness by an action on the reticular activating system

(c) F a marked fall typically occurs due partly to release of vasoconstrictor tone

(d) F body temperature usually falls, perhaps because of the fall in metabolic rate and the peripheral vasodilatation.

571. The figure below shows changes in heart rate (upper line) and in arterial blood pressure (lower line) during a period of forced expiration against a closed glottis (Valsalva manoeuvre) indicated by the dashed vertical lines. In this diagram the:

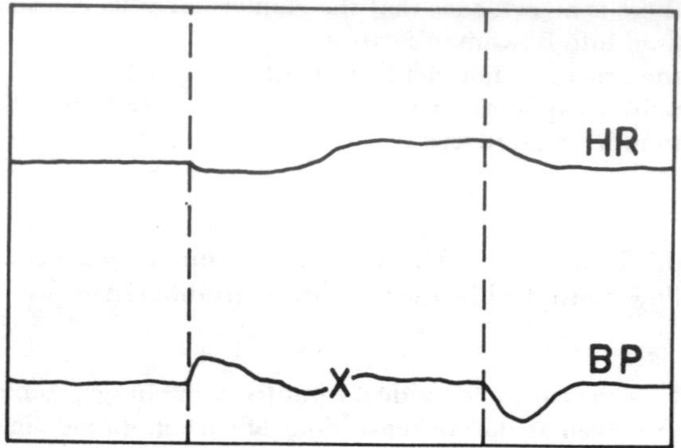

(a) initial rise in blood pressure is due to the baroreceptor reflex

(b) initial rise in blood pressure is likely to coincide with a fall in the venous return to the right atrium
(c) recovery of the blood pressure at X is due to the baroreceptor reflex
(d) changes in the heart rate are brought about by autonomic nerves.

572. The mucosal cells of the intestinal villi resemble the cells of the proximal convoluted tubule of the kidney in that both:
(a) absorb glucose actively
(b) absorb chloride ions actively

(c) absorb amino acids actively
(d) possess microvilli on their absorbing border.

571.

(a) F it is due to transmission to the arterial system of the sudden rise in intrathoracic pressure; the concomitant slowing of the heart suggests that the baroreceptor reflex is *opposing* the rise in pressure

(b) T the sudden rise in intrathoracic pressure would cause a sudden fall in venous return

(c) T it coincides with acceleration of the heart

(d) T the manoeuvre is a way of testing the autonomic nerves to the heart.

572.

(a) T in both cases, absorption is blocked by phloridzin

(b) F in both, the negative chloride ions passively follow the actively absorbed sodium ions

(c) T

(d) T in short, the two types of cell have a very similar function.

573. Exercise which doubles the resting rate of oxygen consumption is likely to about double the:

(a) cardiac output

(b) stroke volume

(c) pressure of carbon dioxide in arterial blood
(d) rate of alveolar ventilation.

574. The endoplasmic reticulum:

(a) is a complex system of tubules in cell nuclei
(b) carries small particles of ribonucleoprotein attached to it

(c) membrane is similar in structure (as seen by the electron microscope) to the membrane surrounding cells
(d) is concerned directly with secretion of sodium ions out of cells.

575. In bone:

(a) osteoclasts are thought to be responsible for bone resorption
(b) a normal calcium content depends on mechanical stress being applied to the bone
(c) the width of the epiphyseal plate is an indication of the rate of growth
(d) strontium ions can replace some of the calcium ions.

573.

(a) T there is a close correlation between oxygen consumption and cardiac output

(b) F the increase in output is effected by an increase in heart rate as well as an increase in stroke volume

(c) F the arterial P_{CO_2} is little changed

(d) T alveolar ventilation is closely related to oxygen consumption in exercise, but the stimulus to ventilation in exercise is little understood.

574.

(a) F it is a tubular system in the cystoplasm

(b) T these particles (ribosomes) are thought to be responsible for the manufacture of proteins such as antibodies, enzymes and hormones

(c) T it may be an extension of the cell membrane in the cytoplasm

(d) F the sodium pump is in the cell membrane.

575.

(a) T they contain an acid phosphatase

(b) T demineralisation occurs when bones are not weight-supporting, e.g. in bed-ridden patients and in astronauts

(c) T it is relatively wide in young individuals

(d) T strontium radioactivity increases the tendency for bone tumours to develop.

576. The standard deviation (SD) of a series of observations:

(a) gives an indication of the scatter of the observations

(b) should be calculated only if the observations have a normal distribution

(c) is a measure of the significance of the observations

(d) may be calculated using the formula:

$$\sqrt{\frac{\Sigma(x+\bar{x})^2}{n}}$$

Σ = sum of; $\bar{x}$ = mean of n observations.

577. Human circadian (24-hour) rhythms:

(a) include changes in body temperature, this reaching a maximum around 4–8 a.m.

(b) include changes in blood growth hormone level, this reaching a maximum around 4–8 a.m.

(c) tend to follow the original time course for some days after the individual has adopted a new daily routine of rest and activity, e.g. going on a night shift

(d) include parallel fluctuations in the blood eosinophil count and in the plasma cortisol level.

578. The membrane which surrounds mammalian cells:

(a) can normally be seen as a dense line in histological sections with the light microscope

(b) consists of a layer of protein sandwiched between two layers of lipid

(c) contains pores which can be seen with the electron microscope

(d) normally contains enzymes.

576.

(a) T on average 66% of observations lie within ±1 SD of the mean; 95% lie within ±2 SD

(b) T otherwise the comment in (a) does not apply

(c) F it is nothing more than a description of the scatter of the observations

(d) F two variations of the formula are:

$$\sqrt{\frac{\Sigma(x-\bar{x})^2}{n}} \quad \text{and} \quad \sqrt{\frac{\Sigma x^2 - \frac{(\Sigma x)^2}{n}}{n}}$$

n − 1 may be substituted for n to give a better estimate of the SD of the population from which the sample (n) is drawn.

577.

(a) F the maximum is in the early evening; the minimum in the early morning

(b) T this may account for the greater rates of cell mitosis and anabolic processes that occur at night

(c) T this accounts also for disturbances in people who fly across time zones

(d) F the fluctuations have a mirror-image relationship since cortisol depresses the eosinophil count.

578.

(a) F although the outline of cells can often be seen, the actual membrane is beyond the power of resolution of the light microscope

(b) F it is thought to consist of lipid with protein structures scattered through it; it also contains carbohydrate

(c) F it acts as if it had pores but no pores can be seen in electron micrographs

(d) T these are involved in the active transport of particles across the membrane.

579. During inspiration:

(a) heart rate tends to decrease

(b) jugular venous pressure tends to fall

(c) the abdominal wall tends to bulge due to relaxation of the abdominal muscles

(d) the intrapulmonary pressure falls below the intrapleural pressure.

580. Lysosomes:

(a) are membrane-bound organelles in the cytoplasm of most cells

(b) contain lysozyme

(c) if broken down release enzymes which cause digestion of the cell contents

(d) in neutrophil granulocytes are responsible for their ability to digest phagocytosed material.

581. When a person changes from the standing to the lying position there is a decrease in:

(a) central venous pressure

(b) the ventilation/perfusion ratio in the apical region of the lungs

(c) the rate of formation of urine

(d) the vital capacity.

582. One mole of calcium ion:

(a) is equivalent to 2 osmoles of calcium ion

(b) has the same mass as 2 equivalents of calcium ion

(c) is the amount of calcium in 1 litre of a normal solution of calcium ions

(d) dissolved in 1 litre of water depresses the freezing point by the same amount ($\pm 10\%$) as one osmole of sodium ion.

579.
(a) F it increases—this "sinus arrhythmia" may be due in part to spread of activity from the respiratory to the cardiovascular centres in the medulla
(b) T because the negative intrathoracic pressure sucks blood into the chest from the great veins
(c) F the abdominal wall is pushed out by the rise in intra-abdominal pressure caused by descent of the diaphragm
(d) F it falls below atmospheric pressure, setting up a pressure gradient between the mouth and the lungs but it does not fall below intrapleural pressure.

580.
(a) T
(b) F this is a bacteriocidal agent found in tears
(c) T release of these enzymes (acid hydrolases) is thought to be responsible for decomposition of tissues after death
(d) T the granules in neutrophil granulocytes are thought to be lysosomes.

581.
(a) F it increases, due to return of blood which has "pooled" in the legs under the influence of gravity
(b) T the ratio is lowest in the most dependent parts of the lungs
(c) F there is an increase in urinary output on lying down which has been attributed to decreased ADH output in response to impulses from volume receptors in intrathoracic blood vessels
(d) T the increased blood content of pulmonary vessels reduces the volume of air the lungs can hold and the pushing up of the diaphragm by abdominal contents tends to impede inspiration.

582.
(a) F osmolality depends on the number of particles; therefore one mole of calcium = 1 osmole of calcium
(b) T an equivalent = molecular weight/valency; since calcium is divalent, one mole of calcium equals 2 equivalents of calcium
(c) F a normal solution contains 1 equivalent per litre not one mole per litre
(d) T one osmole of any substance dissolved in 1 litre of water depresses its freezing point by 1·86°C.

583. The Valsalva manoeuvre (a strong expiratory effort with the glottis closed):

(a) produces a strongly positive intrapleural pressure
(b) aids the expulsion of the baby during labour once the cervix has fully dilated
(c) involves contraction of the diaphragm
(d) is followed by a brief period of hypertension and slowing of the heart.

584. Carbonic anhydrase plays a role in the:

(a) production of HCl by the parietal cells of the stomach

(b) reabsorption of hydrogen ions from the fluid in the renal tubules

(c) passage of CO_2 from the pulmonary capillaries to the alveoli
(d) secretion of bicarbonate by the pancreas.

585. The hydrogen ion concentration in a solution:

(a) of pH 5 is 100 times less than in a solution of pH 7
(b) is inversely related to the hydroxyl ion concentration

(c) is approximately doubled or halved with a pH change of 0·3 units
(d) equals the hydroxyl ion concentration at pH 7.

586. The human cell nucleus:

(a) has a membrane which allows the passage of nucleic acid

(b) contains the genes, each of which is a complex molecule of RNA

(c) may have its genetic material condensed in a darkly staining nucleolus

(d) contains 44 chromosomes in the case of somatic cells.

583.

(a) T due to compression of the gaseous contents of the chest

(b) T this is an essential part of normal labour

(c) F the diaphragm is an inspiratory muscle

(d) T the fall in arterial pressure during the manoeuvre (due to the reduction in venous return) induces reflex vasoconstriction which persists after the manoeuvre ends. The resulting hypertension causes reflex slowing of the heart.

584.

(a) T H_2CO_3 produced from CO_2 and water dissociates to provide hydrogen ions

(b) F but it plays a role in the secretion of hydrogen ions by the tubule cells

(c) T by aiding the conversion of bicarbonate to CO_2

(d) T as in (a) but this time the bicarbonate ions are secreted.

585.

(a) F it is 100 times greater

(b) T the product of the concentration of H^+ and OH^- is constant

(c) T $\log 2 = 0{\cdot}301$

(d) T at pH 7 the solution is neutral.

586.

(a) T messenger RNA conveys information from the nucleus to the protein-synthesising centres

(b) F the genes are portions of the chromosomes (large DNA molecules complexed with protein)

(c) F the nucleolus, frequently seen in growing cells, is a condensation of RNA which is probably synthesised in the area of the nucleolus

(d) F the diploid number is 46; the nuclei of mature germ cells have the haploid number, 23.

587. Athletes tend to differ from untrained individuals by having a greater:

(a) resting cardiac output
(b) resting heart rate
(c) muscular efficiency at high blood lactate levels
(d) maximum oxygen consumption.

588. The total osmotic pressure (OP) of human plasma:

(a) is about 25 mm Hg (3·3 kPa)
(b) is similar to that of 0·9% NaCl solution
(c) is similar to that of 0·9% glucose solution
(d) opposes the tendency of fluid to leave the capillaries.

589. With increasing age a fall tends to occur in:

(a) mean blood pressure
(b) pulse pressure
(c) vital capacity
(d) residual volume.

590. Ingestion of protein:

(a) increases the resting metabolic rate more than does ingestion of similar quantities of fat or carbohydrate
(b) leads to a more acid urine than does ingestion of fat or carbohydrate
(c) yields energy at the rate of 5·4 kilocalories (23 kJ) per gram when metabolised in the body
(d) permits the body to synthesise the essential amino acids.

587.

(a) F there is no appreciable difference
(b) F it may be as low as 40–50; stroke volume is greater
(c) T for a given degree of exertion they also tend to have a smaller rise in blood lactate
(d) T for healthy untrained individuals the maximum is about 3 litres per minute; highly trained skiers have been observed to have maxima of about 6 litres per minute. The maximum rate of oxygen consumption is an index of physical fitness.

588.

(a) F this is the colloid OP; the total of OP is about 5500 mm Hg (700 kPa)
(b) T both have an osmolality of about 290 milliosmoles per litre
(c) F it is similar to that of an approximately 5% glucose solution (osmotically equivalent solutions must contain the same *number* of particles)
(d) F only the colloid OP does this, the capillary wall is freely permeable to non-colloid particles.

589.

(a) F it tends to rise for reasons which are obscure
(b) F the reverse happens due to decreasing elasticity of the aorta
(c) T
(d) F it increases due in part to the loss of elastic recoil in the lungs.

590.

(a) T the specific dynamic action of protein is much greater than that of fat and carbohydrate
(b) T protein has acidic residues such as sulphate and phosphate
(c) F this is the figure for complete combustion; in the body the value is about 4 kilocalories (17 kJ) per gram
(d) F by definition essential amino acids cannot be manufactured in the body.

591. Drinking a litre of water:
(a) increases the arterial level of antidiuretic hormone
(b) reduces the sodium concentration of blood in the arterial system
(c) has a greater effect on the tonicity of portal venous blood than on that of jugular venous blood
(d) decreases the specific gravity of the body.

592. The Golgi apparatus is:
(a) a characteristic feature of muscle spindles
(b) a complex of tubules and vesicles usually sited near the cell nucleus
(c) usually seen in cells with secretory activity
(d) rarely seen in neurones.

593. A buffer system:
(a) is a mixture of a weak acid and its conjugate base or a weak base and its conjugate acid
(b) prevents any change in hydrogen ion concentration when a strong acid is added to the solution
(c) works best when the acid and the base are in approximately equal concentrations
(d) with a pK of 4·2 would be of greater value in the blood than one with a pK of 6·8.

594. Rapid eye movement (REM) sleep differs from non REM sleep in that:
(a) it is associated with dreaming
(b) it is less effective in relieving tiredness due to sleep deprivation
(c) it is associated with EEG waves of greater amplitude
(d) general muscle tone is similar to that in the alert state.

591.
(a) F it is decreased; hence the excess water is excreted
(b) T this is associated with a fall in the osmotic pressure which stimulates the osmoreceptors
(c) T the water is absorbed via the portal vein
(d) F it increases towards 1·000.

592.
(a) F but Golgi *tendon organs* are stretch receptors in muscle tendons
(b) T
(c) T protein hormones and enzymes which are formed in the endoplasmic reticulum are stored in vesicles in the Golgi apparatus
(d) F neurones are secretory cells; transmitter substance is thought to be packaged into vesicles in the Golgi apparatus before transport to the nerve terminals.

593.
(a) T e.g. the carbonic acid-bicarbonate buffer system H_2CO_3 (acid) $\rightleftharpoons H^+ + HCO^-_3$ (conjugate base)
(b) F it does not prevent change but tends to minimise it
(c) T around this point (pK) there is least change in pH for a given addition of acid or base; pH = pK + log [base]/[acid]
(d) F since blood pH is 7·4, the buffer with a pK of 6·8 would be more effective.

594.
(a) T people woken during REM sleep are usually able to recall dreams
(b) F sleep without REM episodes is much less effective in relieving tiredness
(c) F the EEG waves in REM sleep are high frequency, low amplitude waves; in non REM sleep the reverse is true
(d) F general muscle tone is markedly reduced in REM sleep.

595. From the fetal stage, through childhood, adulthood and old age:

(a) the duration of rapid eye movement (REM) sleep as a percentage of total sleep duration falls
(b) the duration of deep (stage 4) sleep as a percentage of total sleep duration rises
(c) total sleeping time falls until adulthood and then rises in old age
(d) brain mass increases until around the time of birth and then declines.

596. Comparing the situation in a normal adult standing in water up to the neck with the situation when standing in air, the person standing in the water:

(a) has the feet subjected to a water pressure nearer 100 than 20 mmHg (nearer 13 than 2·5 kPa)
(b) has a greater proportion of the blood volume in the thorax
(c) has the greater total peripheral resistance
(d) excretes sodium at a higher rate.

595.

(a) T before birth it exceeds 50%, in adulthood it is about 20%, in old age it falls below 10%

(b) F it falls as with REM sleep

(c) F it falls throughout life, being minimal in old age

(d) F the brain continues to grow in childhood, doubling its weight in the first year.

596.

(a) T a column of water 1360 mm deep exerts a pressure of 100 mm Hg (bearing in mind the density of mercury)

(b) T blood is displaced from the limbs

(c) F the counter pressure exerted by the water prevents pooling of blood in the dependent limbs and hence prevents the fall in blood pressure which increases peripheral resistance reflexly when standing in air

(d) T this is associated with a diuresis in response to the increased thoracic blood volume.

GENERAL QUESTIONS
APPLIED PHYSIOLOGY

597. The diagram below shows some lines relating weight to age where B is the normal average for males. In this diagram:

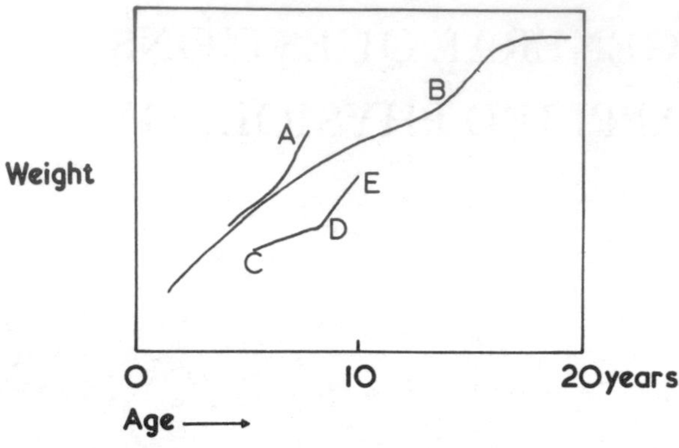

(a) line A would be consistent with precocious puberty

(b) line CDE would be consistent with dwarfism for which appropriate treatment began at point C rather than at point D
(c) the normal line for females would be below but parallel to the line for males
(d) the line for normal weight of lymphoid tissue would show similar trends to line B.

598. After a major operation or serious injury the:
(a) plasma cortisol level rises
(b) body goes into negative nitrogen balance
(c) urinary excretion of potassium rises
(d) tendency for the blood to clot decreases.

599. Obesity is treated successfully by:
(a) taking steam baths daily
(b) adding slimming foods to the normal diet
(c) increasing regular exercise without increasing food intake

(d) eating as much as one likes but avoiding foods with a high fat content.

597.

(a) T a very early growth spurt follows previously normal progression

(b) F it is at point D that the line begins to move towards the normal line

(c) F the female line shows an earlier pubertal growth spurt

(d) F lymphoid mass reaches a maximum around 10–12 years and then declines.

598.

(a) T this aids recovery by mechanisms which are obscure

(b) T cortisol breaks down body protein to form glucose

(c) T again due to adrenal cortical hormones

(d) F it increases, reaching a maximum around 10 days after the trauma.

599.

(a) F this reduces body water but not body fat

(b) F this makes the patient fatter

(c) T this increases energy expenditure without increasing energy intake

(d) F the only treatment is to reduce energy intake relative to energy expenditure. All other treatments are merely aids to this end.

600. An injection of atropine typically produces:

(a) weakness of skeletal muscles throughout the body

(b) dryness of the mouth

(c) decreased secretion of mucus in the airways

(d) blockade of impulse transmission at autonomic ganglia.

601. Oxygen pressure in arterial blood is more affected (as a percentage of normal) than oxygen content by:

(a) carbon monoxide poisoning

(b) treatment with oxygen at 3 atmospheres (300 kPa)

(c) anaemia

(d) a 30% fall in the Po_2 of the inspired air.

602. When 50 g glucose is taken by mouth in a glucose tolerance test:

(a) glucose does not usually appear in the urine if the person is healthy
(b) there will be a greater rise in blood glucose in a diabetic than in a normal person
(c) there will be a slower-than-normal rise in blood glucose in a person with atrophy of the intestinal villi (coeliac disease)
(d) there will be a more rapid return of blood glucose to the resting level in a diabetic than in a normal person.

600.
(a) F atropine does not block the action of acetylcholine at the motor end plate
(b) T by blocking the action of acetylcholine at autonomic nerve endings in the salivary glands
(c) T mucus secretion is mediated by cholinergic nerves; its abolition is the main reason for using atropine before a surgical operation
(d) F though ganglionic transmission is cholinergic, atropine does not block acetylcholine at these sites; drugs like hexamethonium do.

601.
(a) F oxygen *content* is markedly reduced; pressure will be affected only if the Po_2 in the inspired air falls
(b) T the blood carries more oxygen mainly because of the extra dissolved oxygen
(c) F the oxygen pressure is normal since it depends on the Po_2 in inspired air; the oxygen *content* is low since it depends on the haemoglobin concentration
(d) T this reduces the Po_2 in alveolar air and hence the Po_2 in arterial blood; because of the shape of the oxygen dissociation curve there is little change in arterial oxygen content.

602.
(a) T because the normal Tm for glucose is not exceeded
(b) T due to insufficient secretion of insulin
(c) T coeliac disease causes malabsorption which includes slow absorption of glucose
(d) F there is delayed return in patients with diabetes mellitus.

603. An increase above normal in the concentration of circulating red cells (polycythaemia) may be caused by:

(a) lung disease which causes a fall in arterial Po_2

(b) heart disease where there is shunting of blood from the right atrium or ventricle to the left

(c) pregnancy

(d) repeated injections of vitamin B_{12}.

604. Assuming one is kept afloat by a life jacket the survival time in water at 15°C is likely to be:

(a) 12–24 hours for an average unclothed man

(b) extended if the individual is wearing many layers of clothing together with gloves and boots

(c) extended by increasing the metabolic rate by swimming rather than minimising it by floating motionless

(d) more prolonged in fat than in thin persons.

605. An organ transplant in an adult is less likely to be rejected if the recipient:

(a) is given glucocorticoid treatment

(b) has previously received a skin graft from the same individual

(c) receives the organ from his identical twin rather than from his non-identical twin

(d) is of the same blood group as the donor.

606. Obesity:

(a) is associated with a reduced life expectancy

(b) is a recognised cause of diabetes melitus

(c) in parents is associated with obesity in their children

(d) is usually due to a disorder of the endocrine glands.

603.

(a) T the fall in tissue oxygen tension stimulates erythropoietin production

(b) T this results in a low arterial Po_2 and a fall in tissue oxygen tension

(c) F in pregnancy, there is a relatively greater increase in plasma volume than in red cell mass so that the erythrocyte concentration falls ("physiological anaemia")

(d) F though vitamin B_{12} is needed for normal red cell production, it does not regulate red cell production.

604.

(a) F after 1–2 hours the average man becomes helpless due to hypothermia and dies

(b) T this reduces heat loss by trapping layers of warm water next to the body and so reducing the temperature gradient between skin and water

(c) F movements increase heat loss by disturbing the layer of warmed water next to the body

(d) T because of extra insulation; women, having more skin fat, tend to survive longer than men in cold water.

605.

(a) T this tends to suppress the antigen-antibody reactions

(b) F this would be likely to sensitise the recipient and increase the risk of rejection

(c) T only identical twins have identical genes and hence identical tissue antigens

(d) T however, blood group compatibility is only a poor indicator of tissue compatibility; lymphocyte compatibility may be a better guide to the compatibility of the transplantation antigens.

606.

(a) T a person overweight by about 20% has a reduction in life expectancy of about 20%

(b) T mild diabetes in some obese patients may be cured if they lose weight

(c) T fat parents tend to have fat children who grow into fat adults

(d) F obesity is very rarely due to an endocrine disorder.

607. A kidney transplanted into a patient with severe renal failure is:

(a) unlikely to reduce the blood urea level to normal if only one kidney is transplanted

(b) probably being rejected if there is a sharp fall in glomerular filtration rate

(c) probably being rejected if there is a sharp rise in central body temperature in the absence of infection

(d) probably being rejected if urine is small in volume and highly concentrated.

608. A hole suddenly appears in the fuselage of an aircraft at 50 000 ft (15 000 m) where the atmospheric pressure is around 100 mm Hg (13·3 kPa). Likely consequences of loss of cabin pressure include:

(a) rupture of the ear drums

(b) pains in the limbs and a choking sensation

(c) a gradual loss of consciousness after 10–15 minutes

(d) an alveolar oxygen tension of around 80 mm Hg (10·6 kPa) in an individual breathing pure oxygen at the ambient pressure.

609. Helium is often substituted for nitrogen in the gas breathed by divers because it:

(a) is more soluble in body fluids

(b) diffuses through the tissue more rapidly than nitrogen

(c) has less narcotic effect than nitrogen

(d) is more viscous than nitrogen.

607.
(a) F one kidney is adequate provided it is normal

(b) T this suggests nephron damage

(c) T this, like a raised ESR and granulocyte count, is an accompaniment of tissue destruction

(d) F a highly concentrated urine suggests good renal function.

608.

(a) T because atmospheric pressure drops much more suddenly than can middle ear pressure; the ear drums burst outwards

(b) T these are symptoms of decompression sickness which is caused by nitrogen bubbles suddenly coming out of solution and damaging the tissues

(c) F even if the sudden decompression did not cause unconsciousness, consciousness would be lost in less than a minute due to lack of oxygen

(d) F of the total alveolar tension more than half will be taken up by CO_2 and water vapour so that consciousness will still be lost.

609.

(a) F it is less soluble so that less goes into solution during the dive. There is less danger of its bubbling out of solution during ascent to cause decompression sickness

(b) T this is so because it has a smaller molecule, therefore, less time is needed for decompression

(c) T when nitrogen goes into solution during diving, N_2 narcosis (confusion, loss of muscle power and eventually unconsciousness) occurs. Helium is less narcotic

(d) F it is less viscous; this diminishes the work required to shift the compressed gas while breathing under pressure.

610. An individual who has just received an electric shock which has produced ventricular fibrillation:

(a) is likely to lose consciousness after 5–10 minutes
(b) will have a weak carotid pulse at around 200–250 beats per minute
(c) should be given external cardiac massage immediately he has been removed from the electrical contact
(d) need not be given artificial ventilation until after the ventricular fibrillation has been reversed since it would serve no purpose until the heart's action is normal.

611. The restoration of the blood volume after haemorrhage is aided by:

(a) the mobilisation of fluid from the intracellular fluid compartment
(b) an increase in the osmotic pressure of the plasma proteins
(c) contraction of venous reservoirs
(d) arterial vasoconstriction.

612. A patient suffering from a feverish illness:

(a) tends to feel coldest while his central body temperature is rising
(b) shows marked body temperature swings with swings in environmental temperature
(c) has a raised basal metabolic rate
(d) sweats only when his central body temperature is falling.

613. Acidosis is associated with:

(a) a lowered blood potassium level (hypokalaemia)
(b) chronic renal failure
(c) severe diarrhoea
(d) hypoventilation.

610.

(a) F consciousness is lost in less than a minute
(b) F there is no appreciable cardiac output; hence no pulse

(c) T this will maintain an adequate output until the fibrillation is reversed
(d) F artificial ventilation should accompany cardiac massage which maintains some circulation; mouth-to-mouth is the most efficient method.

611.

(a) F this does not occur since haemorrhage does not alter the tonicity of the extracellular fluid
(b) F if anything, plasma protein concentration falls after haemorrhage as fluid moves from the interstitial fluid to the blood
(c) F this redistributes but does not increase blood volume
(d) T the fall in blood pressure causes reflex arteriolar vasoconstriction. This lowers capillary blood pressure so that there is net fluid gain (by osmosis) by the capillaries.

612.
(a) T activity of the heat-regulating centre leads to heat-conserving phenomena; these are associated with cutaneous vasoconstriction and the sensation of feeling cold
(b) F central temperature is still regulated precisely but at a raised level
(c) T the raised tissue temperature speeds metabolism
(d) F sweating may occur if body temperature rises above the new set level.

613.
(a) F potassium competes with hydrogen for secretion in exchange for sodium in the renal tubules; the blood potassium level tends to rise
(b) T due to the failure of the kidney to excrete the acid residues of protein digestion
(c) T the intestinal contents are rich in bicarbonate so a metabolic acidosis results
(d) T CO_2 retention causes a respiratory acidosis.

614. When a disease is inherited as a recessive autosomal character:

(a) one of the parents of the patient will exhibit the disease
(b) all the children of the patient will exhibit the disease

(c) both parents of the patient carry the recessive character

(d) subsequent brothers or sisters of the patient have a one in two chance of inheriting the disease if both parents are heterozygous.

615. A report comparing two forms of treatment (A and B) states that treatment A was superior in certain respects ($P<0{\cdot}01$). It can be concluded that:

(a) the chances of being improved by treatment A are more than 99% in a given case

(b) the observed superiority is statistically significant

(c) at least 100 patients were studied
(d) treatment A should be substituted in future for treatment B.

616. A low serum potassium level:

(a) is associated with repeated vomiting of gastric contents

(b) causes impairment of intestinal peristalsis

(c) is caused by aldosterone deficiency
(d) indicates that there is depletion of the body potassium content.

617. A metabolic acidosis (e.g. diabetic ketosis) is typically associated with:

(a) a blood pH of less than 6·5
(b) a rise in the capillary P_{CO_2} level
(c) a urine pH of less than 6·5

(d) tetany.

614.
(a) F both parents are usually healthy
(b) F the children will probably be normal carriers since the recessive genetic character will be masked by the normal gene donated by the spouse
(c) T both parents must have contributed an abnormal gene to form a homozygous genotype
(d) F from simple Mendelian laws the chances are one in four.

615.
(a) F $P<0{\cdot}01$ means that the likelihood of the observed difference between treatments A and B being due to chance is less than 1 %
(b) T conventionally a P value less than 0·05 (one in twenty) is taken as indicating a significant difference between groups of results
(c) F the P value gives no indication of the number of patients
(d) F the superiority, though statistically significant, may be slight, and adverse effects could outweigh the benefits.

616.
(a) T due to the cast-off cells lost in the vomitus and the effect of alkalosis on the renal tubular secretion of potassium
(b) T potassium depletion interferes with the working of intestinal smooth muscle and may cause paralytic ileus
(c) F aldosterone favours potassium excretion by the kidney
(d) F significant depletion of the serum potassium may occur even when the total potassium content of the body is increased.

617.
(a) F death is likely to occur before the pH falls to this level
(b) F P_{CO_2} falls below the normal value due to hyperventilation
(c) T urinary pH is likely to be acid, e.g. around 5·0 unless the acidosis is due to excessive renal excretion of bicarbonate
(d) F acidosis reduces the ability of plasma proteins to bind calcium so that the ionised calcium concentration rises.

618. When a disease is inherited as a sex-linked recessive genetic character:
(a) the gene is located on the Y chromosome
(b) the disease will be seen more commonly in males than in females
(c) the disease can be transmitted by the female but not the male
(d) it may fail to manifest itself in a female carrier.

619. When the osmolality of the extracellular fluids rises:
(a) the subject experiences thirst
(b) reabsorption of water by the proximal convoluted tubules is increased
(c) the intracellular fluid volume falls
(d) sweat secretion stops.

620. During severe exercise there is usually an increase in:
(a) arterial blood pressure
(b) blood lactic acid
(c) tissue fluid formation in the active muscles
(d) urinary output.

621. When a person moves from a temperate to a tropical climate:
(a) his basal metabolic rate falls
(b) his cardiac output rises
(c) his ability to regulate body temperature improves as he adapts to the new climate
(d) he is more likely to experience muscle cramps.

622. When the respiratory tract is suddenly and completely obstructed:
(a) respiratory efforts increase
(b) the pupils dilate
(c) consciousness is lost immediately
(d) the patient may scream.

618.

(a) F the gene is located on the X chromosome
(b) T because the heterozygous female will be protected by the other X chromosome
(c) F both the male and the female can transmit the X chromosome
(d) T see (b); haemophilia and Christmas disease are inherited as sex-linked recessive characters.

619.

(a) T probably a direct action on the thirst centre in the hypothalamus
(b) F reabsorption of water by the proximal convoluted tubules is not regulated to meet body water requirements
(c) T water moves out of the cells due to osmotic forces
(d) F sweat secretion is not related to the salt and water needs of the body.

620.

(a) T the rise in cardiac output is proportionately greater than the fall in peripheral resistance
(b) T the result of anaerobic metabolism
(c) T due to the rise in capillary blood pressure; increased lymph flow compensates for this
(d) F reflex vasoconstriction in the kidney and increased ADH release decrease urinary output.

621.

(a) T due to diminished thyroid gland activity; the hypothalamus releases less TRH
(b) T due in part to the increased circulation through the skin
(c) T after adaptation the ability of the subject to secrete sweat is increased
(d) T due to loss of salt in sweat.

622.

(a) T due to CO_2 accumulation and oxygen deficiency acting on the central and peripheral chemoreceptors
(b) T due to increased sympathetic activity
(c) F consciousness is retained for about 30–60 seconds
(d) F air movement past the larynx is necessary for screaming.

623. An adult differs from an infant in that his:

(a) nitrogen balance is usually positive

(b) extracellular fluid is a larger proportion of his body weight

(c) blood contains reticulocytes

(d) homeostatic mechanisms are, in general, more efficient.

624. When a disease is inherited as a dominant autosomal character:

(a) the disease is usually severe enough to prevent reproduction

(b) males and females are equally affected

(c) all the children of the affected adult must exhibit the disease

(d) one of the parents of the patient must be a carrier without signs of the disease.

625. Excessive sweating (hyperidrosis) in a limb is likely to be relieved by:

(a) ganglion-blocking drugs such as hexamethonium

(b) a drug which interferes with transmission at cholinergic nerve endings

(c) a drug which interferes with transmission at adrenergic nerve endings

(d) division of the parasympathetic motor neurones supplying the region.

623.
(a) F adults are normally in nitrogen balance, infants are in positive nitrogen balance
(b) F in infants, extracellular fluid exceeds intracellular fluid; the intracellular compartment becomes the larger during childhood
(c) F both adults and infants have reticulocytes in their blood
(d) T adults can tolerate environmental stress better than infants.

624.
(a) F if this were so, the genes for the disease would rapidly disappear from the population
(b) T
(c) F only half of the children will exhibit the disease if the affected adult is heterozygous and marries a normal person
(d) F if the character is dominant all carriers will exhibit the disease.

625.
(a) T but side-effects are likely to be worse than the disease
(b) T e.g. atropine; but again side-effects may be troublesome
(c) F the sympathetic nerves supplying sweat glands are cholinergic
(d) F there are no parasympathetic fibres in the limbs; division of sympathetic neurones is usually effective.

626. A person who has not yet adapted to residence at an altitude of 6 000 metres, where the atmospheric pressure is half that at sea level, experiences mountain sickness and:

(a) has a rate of alveolar ventilation similar to that at sea level

(b) has a reduced exercise tolerance due mainly to respiratory alkalosis

(c) runs the hazard of developing pulmonary oedema

(d) his symptoms may be alleviated by drugs which inhibit the action of carbonic anhydrase.

627. Hiccuping:

(a) necessarily includes a sudden closure of the glottis
(b) necessarily includes an involuntary inspiration

(c) is a reflex phenomenon

(d) can be reduced in severity by blocking conduction in one of the phrenic nerves.

628. In the first few days following a major injury or surgical operation:

(a) there is increased glucose formation from non-carbohydrate sources
(b) the patient is in nitrogen balance

(c) potassium balance tends to be negative because of diminished potassium intake in the diet

(d) protoplasmic protrusions can be seen projecting from the surfaces of blood platelets.

626.

(a) F his alveolar ventilation is increased due to hypoxic stimulation of his peripheral chemoreceptors

(b) F the hypoxic hypoxia is the main reason for the reduced exercise tolerance. The low P_{CO_2} and hydrogen ion concentration would tend to extend exercise tolerance rather than limit it

(c) T this can be a fatal hazard of mountaineering but the mechanism is obscure; such patients may be helped by oxygen therapy and by bringing them quickly to a lower altitude

(d) T carbonic anhydrase inhibitors cause an increased excretion of bicarbonate in the urine which, by raising the ratio of P_{CO_2}: (HCO_3^-) towards the normal (1:20) helps to compensate for the metabolic alkalosis.

627.

(a) T this produces the characteristic clicking sound

(b) T this brief inspiration is terminated by closure of the glottis

(c) T it is essentially a reflex initiated by irritation in the region of the diaphragm; metabolic abnormalities, such as uraemia can also cause it

(d) T the involuntary inspiration is due to spasmodic contraction of the diaphragm; persistent hiccuping in a seriously ill patient may cause considerable distress.

628.

(a) T the increase in gluconeogenesis is attributed to increased secretion of cortisol

(b) F the breakdown of protein for gluconeogenesis causes a negative nitrogen balance

(c) F the negative balance is due to cortisol-induced secretion of potassium by the renal tubules and the release of potassium from autolysed cells

(d) T the spikes increase the tendency of the platelets to adhere together to form clumps; the tendency for blood to clot increases.

629. A person breathing oxygen at a pressure of four atmospheres (about 400 kPa) for one hour:

(a) would have an increased oxygen consumption if his metabolic rate remained constant

(b) is likely to show changes in his nervous system which result in pins and needles sensations, irritability, and muscle twitches, culminating in convulsions

(c) has a decrease in cerebral blood flow

(d) shows a decreased sensitivity to ionising radiation.

630. The graph shows the heart rate of someone who received painful dental treatment at point C. In this patient:

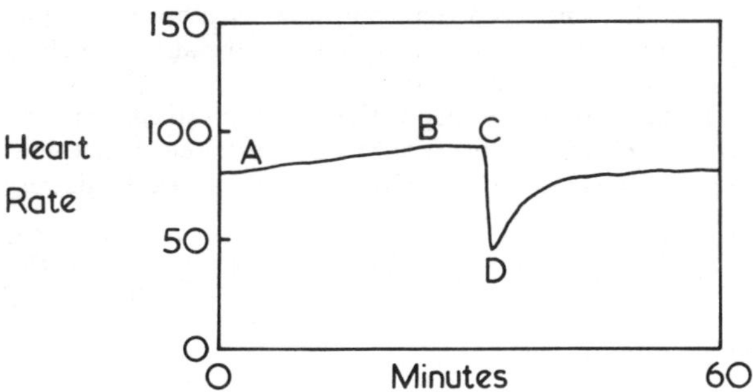

(a) the blood catecholamine level is likely to have been higher at point B than point A

(b) the sudden fall in heart rate (CD) is likely to have been due to an abrupt fall in the blood catecholamine level

(c) vagal tone is likely to have reached a maximum at point D

(d) blood pressure is likely to have fallen rather than risen between C and D.

629.

(a) F oxygen consumption is determined by the metabolic rate, not the Po_2 in the inspired air

(b) T these are the symptoms of *oxygen toxicity* whose mechanism is obscure. It has been attributed to tissue damage by oxidising radicals formed at the high oxygen pressures

(c) T oxygen lack is a dilator and oxygen excess a constrictor of cerebral blood vessels

(d) F sensitivity to ionising radiation is increased, perhaps because oxidising radicals are more easily formed by ionisation in these conditions. Hyperbaric oxygen is sometimes used to increase the sensitivity of cancer tissues in radiotherapy.

630.

(a) T anticipation of the treatment seems to have produced an increasing level of sympathetic stimulation

(b) F breakdown of catecholamines is relatively slow and the level cannot fall dramatically within a minute or two

(c) T a sudden rise in vagal tone (caused by pain) would explain the sudden fall in heart rate

(d) T pain tends to produce hypotension by both vagal slowing of the heart and by vasodilatation due to cholinergic sympathetic discharge (vasovagal fainting).

631. The ratio of intravascular hydrostatic pressure to colloid osmotic pressure is greater:

(a) in splanchnic than in renal glomerular capillaries

(b) than normal in the circulation of someone suffering from hepatic failure

(c) than normal in the systemic capillaries of someone who has just suffered a severe blood loss

(d) than normal in capillaries in a region where there is oedema due to venous obstruction.

632. Sudden application of cold water to the:

(a) hand tends to cause an increase rather than a decrease in blood pressure

(b) face tends to cause an increase rather than a decrease in the heart rate

(c) oesophagus (by drinking a considerable amount) can produce changes in the electrocardiogram

(d) external auditory meatus is likely to produce both nausea and nystagmus.

633. A normal healthy young man can withstand loss of half of his:

(a) renal function without developing renal failure

(b) pulmonary function without developing respiratory failure

(c) circulating platelets without developing a haemorrhagic tendency

(d) seminal sperm count without suffering from infertility.

631.

(a) F it is higher in glomerular capillaries, as is necessary for filtration

(b) T the liver is the source of circulating albumin which accounts for the greater part of the colloid osmotic pressure

(c) F the reverse is true; the relatively great osmotic pressure in the systemic capillaries draws in tissue fluid and helps to restore the blood volume

(d) T the relatively great hydrostatic pressure forces fluid out of the capillaries into the interstitial spaces.

632.

(a) T this is the cold pressor response

(b) F the typical response is a bradycardia (this is known as the diving reflex; in some mammals, e.g. seals, diving causes extreme bradycardia)

(c) T the myocardium is thereby cooled

(d) T the vestibular apparatus is cooled; convection currents produce intense stimulation of the hair cells.

633.

(a) T normal function can be maintained with one kidney

(b) T with one healthy lung considerable exertion is still possible

(c) T the platelet count must fall by more than half before haemorrhagic problems arise

(d) T in short, most body functions carry at least a 50% reserve in the young adult.